Alexandre Tesla

Biodiesel production101

Homebrew Edition

A do it yourself guide to produce biodiesel on your backyard

© Copyright 2018 Alan Adrian Delfin Cota- All rights reserved.

This document is geared towards providing exact and reliable information in regards to the topic and issue covered. The publication is sold with the idea that the publisher is not required to render accounting, officially permitted, or otherwise, qualified services. If advice is necessary, legal or professional, a practiced individual in the profession should be ordered.

- From a Declaration of Principles which was accepted and approved equally by a Committee of the American Bar Association and a Committee of Publishers and Associations.

In no way is it legal to reproduce, duplicate, or transmit any part of this document in either electronic means or in printed format. Recording of this publication is strictly prohibited and any storage of this document is not allowed unless with written permission from the publisher. All rights reserved.

The information provided herein is stated to be truthful and consistent, in that any liability, in terms of inattention or otherwise, by any usage or abuse of any policies, processes, or directions contained within is the solitary and utter responsibility of the recipient reader. Under no circumstances will any legal responsibility or blame be held against the publisher for any reparation, damages, or monetary loss due to the information herein, either directly or indirectly.

Respective authors own all copyrights not held by the publisher.

The information herein is offered for informational purposes solely, and is universal as so. The presentation of the information is without contract or any type of guarantee assurance.

The trademarks that are used are without any consent, and the publication of the trademark is without permission or backing by the trademark owner. All trademarks and brands within this book are for clarifying purposes only and are the owned by the owners themselves, not affiliated with this document.

Disclaimer and Terms of Uses

The AuthorandPublisher has strived to be as accurate and complete as possible in the creation ofthis book, notwithstandingthe fact that he does notwarrantor represent at any timethatthe contents within are accurate due to the rapidly changing nature of the Internet. Whileall attempts have been made to verify information provided in this publication, the AuthorandPublisher assumes no responsibility for errors, omissions, or contrary interpretation of the subject matter herein.

Acknowledgement

I have taken effort in this project. However it would not have been possible without the kind support and help of many individuals who contribute on this project, especially Octavio Peláez, who played a key role to write this book...

ABSTRACT

This document provides a general description of alternative fuels developed around the world to meet your needs. Scientists are trying to find new fuels that can solve the sources of depletion. Biofuels such as butanol, methanol, ethanol, biogas, biodiesel, etc. are being tested. Tests are underway to convert natural crop plants into an environmentally friendly form of diesel, but the differences in costs make this area difficult, but as fossil fuels begin to deplete, this is an option that can be pursued. Here we discuss the types of fuels used by all inhabitants and try to find a solution to the problem of reducing fossil fuels, that is, the crisis of conventional fuels. The risks associated with its use are discussed and the benefits are listed. Biofuels can be used for centralized electricity, district heating, and local heating.

A Do it yourself guide to diesel, black, diesel, red diesel, diesel non-road, marine diesel, kerosene & liquefied natural gas

Contents

- INTRODUCTION 10
  - 1.1 What is biodiesel? 12
  - 1.2 Biodiesel Reaction 13
  - 1.3 Why produce my own biodiesel? 15
  - 1.4 Why is not sold everywhere? 17
  - 1.5 Downsides and benefits of using biodiesel 21
    - 1.5.1 Benefits of Biodiesel 21
    - 1.5.2 Disadvantages of biofuels: 23
  - 1.6 Sustainable of biodiesel 24
- Introduction 25
  - 2.1 HOW DO DIESEL ENGINES WORK? 27
  - 2.2 DIESEL ENGINE 30
  - 2.3 The basic four-stroke diesel engine 31
  - 2.4 The two-stroke diesel 32
  - 2.5 Types of Diesel Engine 32
  - 2.6 Why You Should Consider Choosing a Diesel Engine Vehicle 33
  - 2.7 Advantages of Diesel Engines 34
  - 2.8 Diesel Engine Problems 35
  - 2.9 Why diesel engines are still popular? 37
  - 2.10 Benefits of Diesel Engines 37
- The raw material used in production of biodiesel 39
  - 3.1 Introduction to Biodiesel Production 40
  - 3.2 Edible vegetable oils 40
  - 3.3 Inedible vegetable oils 41
  - 3.4 Used Edible Oils 42
  - 3.5 Microalgae 43
  - 3.6 Animal Fats 43
  - 3.7 Materials 44
    - 3.7.1 Materials handling: Oil 44
    - 3.7.2 Materials handling: Methanol 44

3.7.3 Materials handling: Catalyst 44

3.7.4 Waste stream handling 45

3.8 How to transport the material 45

3.9 Health and safety 46

3.9.1 Definitions 50

3.9.2 Fire Extinguishers 51

3.10 Considerations Regarding Methanol 54

3.10.1 EMERGENCY OVERVIEW 55

3.10.2 FIRST AID MEASURES 56

3.11 Physical and Chemical Properties of methanol 57

3.12 The Dangers of Oily Rags 58

3.13 Hazardous Material Classifications and Spills 59

3.14 Hazards Involving Biodiesel Production 60

3.15 Biodiesel production permitting considerations 61

3.15.1 Oil collection 62

3.15.2 Zoning 62

3.15.3 Air quality and construction permits 62

3.15.4 Operating permit? 63

3.16 Glycerol 64

3.17 EPA and state regulation of storage tanks 65

3.18 Dispensing issues 65

3.19 Workplace safety 65

3.20 Discharge to sanitary sewer 67

3.21 Blending and storage of biodiesel 68

3.22 Storage Tanks and Dispensing Equipment 68

3.23 Materials Compatibility 69

3.24 Cleaning Effect 70

3.25 Microbial Contamination 71

3.26 Codes and Regulations 71

Design and construction of a small-scale biodiesel plant 74

4.1 Process Design 74

4.2 Preheat Stage 75

4.3 Processing Stage 77

4.4 Washing Stage 78

4.5 Methanol Recapture 80

4.6 Water Treatment 82

4.7 Outline of Water Recycling Process 82

    4.7.1 Solar Distillation 83

    4.7.2 Slow Sand Filtration 83

    4.7.3 Methanol Evaporation 83

    4.7.4 Solar Heating 84

4.8 Obstacles and Problem Solving 84

    4.8.1 Technical obstacles 85

    4.8.2 Equipment Obstacles 88

5.0 Base Transesterification 93

5.1 Acid Esterification 98

5.2 Expected Yields 100

5.3 Building an Appleseed Processor 103

    5.3.1 Diagrammed Wash tank 112

    5.3.2 Simplified Appleseed Processor 115

    5.3.3 Waste-Oil Funnel 115

6.0 Introduction 120

6.2 Chemical reaction for biodiesel production – transesterification 122

6.3 Making Biodiesel 124

    6.3.1 Collecting oil 124

    6.3.2 Filtering and settling the oil 127

    6.3.3 Migrating the oil to the processor 131

    6.3.4 Evaluating the quality of the mixed oil 135

    6.4.5 Determine the recipe for the methoxide 138

    6.4.6 Warm the oil 141

    6.4.7 Make the methoxide 144

    6.4.8 Drain the glycerin 150

6.4.9 Pump the biodiesel into your standpipe wash tank 154

6.4.15 Drain and dry the biodiesel 165

6.4.16 Filter the biodiesel 170

6.4.17 Fill your tank 173

6.5 Safety issues during processing 173

1. Biodiesel

INTRODUCTION

Fossil fuels provide the bulk of the energy consumed today. But the continued use of these fossil fuels leads to their depletion. It is possible that the world is heading towards a global energy crisis due to a decrease in the availability of cheap oil and recommendations to reduce dependence on fossil fuels. This has led to a growing interest in research on alternative energies/fuels, such as fuel cell technology, hydrogen, bio methanol, biodiesel, solar energy, geothermal energy.

Biodiesel is a renewable domestic fuel for diesel engines derived from natural oils such as vegetable oils. Biodiesel can be used in any concentration with petroleum-based diesel fuel in existing diesel engines with little or no modification. Biodiesel is not the same as raw vegetable oils. It is produced by a chemical process that removes glycerol from the vegetable oil. Chemically, a mixture of monoalkyl esters of long chain fatty acids is a fuel. Unlike pure vegetable oils, biodiesel has very similar uses.

However, most of the time, it is used as an additive for petroleum-based diesel, thus improving the low lubricity of pure and ultra-low sulphur petroleum diesel fuel. It is one of the potential candidates to replace fossil fuels as the world's leading source of transportation, energy since it is a renewable fuel that can replace petrol, diesel in existing engines and can be transported and sold using current infrastructure. Biodiesel is usually produced by reacting a vegetable oil or animal fat with an alcohol such as methanol or ethanol in the presence of a catalyst (acid or base) to produce monoalkyl esters and glycerol, which are then eliminated.

The scientists E. Duffy and J. Patrick proceeded with the TRANSESTERIFICATION of a vegetable oil since 1853, many years before the first diesel engine began to operate. The primary model of Rudolf Diesel, a simple 10 foot (3 m) iron cylinder with a steering wheel at its base, was operating for the first time in Augsburg, Germany. This engine was an example of diesel vision because it was fuelled by peanut oil, a biofuel, although not strictly biodiesel, as it was not Trans esterified. He felt that the use of biomass was the actual future of his engine. In the 1920s, diesel engine manufacturers switched their engines to use the lower viscosity of fossil fuel (PetroDiesel) instead of vegetable oil, a fuel derived from biomass.

The oil industries were able to penetrate the fuel markets because their fuel was much cheaper to produce than alternatives to biomass. The result has been, for many years, a virtual elimination of the biofuel production infrastructure. It is only recently that concerns about environmental impact and the reduction of differential costs have made biofuels as biodiesel a growing alternative. In the 1990s, France launched the local production of biodiesel (locally called disaster) obtained by Trans esterification of rapeseed oil. It is mixed with the proportion of 5% in regular diesel fuel and the proportion of 30% in the diesel fuel used by some captive fleets (public transport). Renault, Peugeot, and other manufacturers have certified truck engines that can be used up to this partial biodiesel. Experiments with 50% biodiesel are underway. From 1978 to 1996, the National Renewable Energy Laboratory of the United States experimented with the use of algae as a source of biofuel under the "Aquatic Species Program".

CONCLUSION: Currently, most biofuels burn to release their stored chemical energy. The search for more efficient ways of converting biofuels and other fuels into electricity using fuel cells is a very active area of work. Bioenergy covers about 15% of global energy consumption. Most Bioenergy is consumed in developing countries and is used for direct heating, unlike electricity generation. However, Sweden and Finland provide respectively 17% and 19% of their energy needs in Bioenergy, which is quite high for the industrialized countries. Biomass can be used for both centralized district heat and power generation and local heating.

1.1 What is biodiesel?

Biodiesel is defined as a fuel composed of mono-alkyl esters of long-chain fatty acids derived from vegetable oils or animal fats. Biodiesel is generally created by the reaction of fatty acids with an alcohol in the presence of a catalyst to produce the desired monoalkyl esters and glycerin. After the reaction, the glycerin, the catalyst, and any remaining alcohol or fatty acid are removed from the mixture. The alcohol used in the reaction is usually methanol, although ethanol and higher alcohols have also been used. Most biodiesel currently produced in the United States. It is made from soybean oil and soy biodiesel usually consists of the five methyl esters, whereas pure biodiesel (i.e. 100%) can be used, a mixture of between 2 and 20% (by volume). It is recommended to use biodiesel with diesel fuel to avoid compatibility problems with the engine.

1.2 Biodiesel Reaction

Reacting / In the presence of catalyst / Yields

100 lbs.   Sodium hydroxide or 100 lbs.

Vegetable oil, Potassium hydroxide   Biodiesel

   Or +

Glycerine n

Animal fat

   +

100 Lbs.

Alcohol

Methanol or Ethanol

Transesterification process produces mono-alkyl esters – chemically similar to diesel fuel

This alternative fuel of clean combustion is produced from domestic resources, such as soy, which is entirely renewable. While biodiesel does not contain oil, it can be mixed with oil to create a potent blend of biodiesel. However, biodiesel can be used in diesel engines without any modification, which makes it the best form of biodegradable and non-toxic fuel available. It is so safe that this common table salt has proven to be more toxic than biodiesel.

Unlike conventional vegetable-based fuels, which can only be used in specially designed, modified combustion ignition engines, biodiesel can be used in its purest form to power diesel engines every day. Imagine that companies ship their products on large platforms with biodiesel and farmers use their equipment with this biodegradable product.

The pure chemical process of esterification manufactures biodiesel. During this process, the glycerin is extracted from vegetable fat or oil. The process naturally leaves byproducts, including methyl esters, which is the chemical name of biodiesel and glycerin, commonly used in the production of soap. The best part of biodiesel is that it does not contain sugar or aromas, which is not possible for traditional fuels.

Biodiesel will not even look like other alternative fuels available today. This is the only alternative fuel that has met all the health effects testing requirements of the Amendments to the Clean Air Act of 1990. This means that biodiesel is legally registered with the Environmental Protection Agency as a legal fuel for sale and distribution. Companies that produce alternative vegetable oil cannot sell their products as legal motorized fuel because they cannot meet the fuel specifications required for EPA registration.

Biodiesel is made from renewable resources, which makes it an intelligent fuel option, which guarantees the protection of our environment for future generations. Their emissions are also significantly lower than those of diesel oil that most people install in their vehicles.

Biodiesel is also better for the economy because it is manufactured in the United States from available resources within the borders of those countries. When biodiesel is produced from products grown in the United States, such as soybeans, it helps to understand countries' dependence on foreign oil and reinvests US money in the United States economy.

This innovative fuel is increasingly available. It can be found throughout the country in selected locations or purchased directly from producers and traders. It costs a little more than traditional fuels, but as the demand for biodegradable and safe alternative fuels increases, the price of biodiesel is expected to fall rapidly. The cheapest way to buy biodiesel is to do it yourself at home.

1.3 Why produce my own biodiesel?

As the smartest technologies continue to offer benefits, this includes a positive impact on the environment and an improvement in the economy. Biodiesel is not intended to replace fossil fuels, but to help formulate a balanced policy. Biofuel is created to be an ideal substitute for extending the use of diesel longevity.

The primary objective of biodiesel is to ensure energy security and benefit local communities. We need biofuels for these following reasons.

- **Easy to use**

One of the main reasons we need biofuels is that they can be used in today's engines, infrastructure, and vehicles without the need for modifications. Biofuel can be stored, burned and pumped in the same way as diesel. It can also be used in mixed or pure form safely. Fuel economy will benefit from the use of biofuels because they are almost identical to petroleum and can be used all year round.

While vehicles older than 15 years old will have to be replaced by fuel lines, the biofuel can crack them. Biofuel can release fuel tanks. The user can switch between biofuel and oil if necessary without complications.

- **Biofuel provides energy security**

Energy security is the most consistent supply available and affordable for consumers as well as for the industry. Some of the many risks to energy security, disrupt the supply of fossil fuels, rising energy prices and limited sources of fuels.

At present, many countries are attracted to the idea of using biofuels from local sources to replace them. This mainly concerns the United Kingdom, which now depends on fossil fuels.

- **Build economic development**

Increased investment in biofuels will result in increased economic growth. This means there will be more jobs and new sources of income for farmers in the industry. Developing countries will benefit from the economic growth of global energy demand.

Demand is expected to increase by 84% as new energy sources such as biofuel will meet this requirement.

- **Reduction of greenhouse gases and emissions**

In the United Kingdom, transport generates more than a fifth of total greenhouse gas emissions. With the right production method, biofuels will produce more greenhouse gases than those currently produced from fossil fuels. This opens the opportunity to face the crucial challenges we face today concerning quality and fuel emissions.

Biodiesel is also considered the most effective alternative fuel to complete the stringent emissions and health study conducted under the Clean Air Act by the EPA. Biofuels will reduce the emissions of carcinogenic compounds by up to 85%.

- **Energy balance**

The fuel energy balance is the ratio between the amount of energy required for production, manufacture, and distribution, compared to the amount of energy released during combustion.

As energy security continues to become a hot topic in society and governments around the world, biofuel has a high energy balance compared to other fuels. Without a constant supply of affordable energy, the country's economy will stop without power to run power plants, transport, and heat homes.

Biofuels can help improve energy security and improve the energy balance through national energy crops. The plants are used to produce biofuels instead of imported crude oil. Biofuels will also strengthen the overall national capacity to reduce oil import requirements.

- **Recyclable and biodegradable**

It has been shown that biofuel is less toxic than diesel because its attributes make it less likely to harm the environment and the cost of damage is lower. Biofuels are less toxic than table salt because it is a natural and non-toxic vegetable oil. Biofuels are also less toxic than common fish species.

Biofuels have proven to be safer to handle than petroleum fuels because of their low volatility. A large amount of energy makes it an accidental ignition hazard because the fuel will create enough steam to ignite. It can be made with different oils and greases, including used cooking oils. Recycled oils will increase their value and make them more profitable.

1.4 Why is not sold everywhere?

It seems unlikely that biofuels are widely used or play an essential role in reducing net CO_2 emissions or replacing fossil fuels burned by internal combustion engines, for the following reasons:

• They are a half-way full half of limited use of existing engines;

• An alternative, more efficient and cheaper fuels are available for vehicle propulsion.

• Most internal combustion engines must be replaced mainly within 10 years;

• Biofuels produced in large quantities to create distortions in the market and food shortages;

Their production can damage the environment and the habitat. The majority of currently used vehicle engines cannot use ethanol alone or, in some cases, even as additives to gasoline. Even a mixture of gasoline containing 10% ethanol can cause corrosion of fuel lines and pumps of some vehicles. It is true that these can be replaced by materials insensitive to the corrosive effects of ethanol, which could be done if it were considered a cost-effective measure. But is not the case.

The price difference between gasoline and a 10% blend of ethanol and gasoline generally does not exceed 2 ¢ per litre. This is insufficient to justify the cost of changing to corrosion resistant materials. The increasing number of vehicles on the road, combined with a finite and decreasing supply of underground oil, means that prices of diesel and gasoline will continue to rise.

If they are not entirely or partially replaced, the price of these fuels will not be affordable, especially for car trips, in the next three to five years. The widespread availability of alternative fuel, electricity, suggests that electric cars will be produced and used in increasing amounts, especially for daily commuting in and around major urban areas. Electricity is much more efficient and much cheaper to use than gasoline or diesel, even at current prices.

Stimulated by the need to reduce CO_2 emissions and the increased cost of fossil fuels, it is expected that the electric motor will replace the internal combustion engine in ten years. There are also significant economic and political reasons for this to happen.

Battery technology has recently made significant progress in enabling the production of smaller, lighter and more durable batteries that can maintain a much higher charge, ten times that of conventional vehicle batteries. It is important to note that these batteries can be recharged quickly, in minutes instead of hours in the case of vehicle batteries and can be manufactured more economically.

Ongoing research by CSIRO, MIT, and other organizations increases the likelihood of further improvements over the next three to five years. These improvements will encourage the development and increased use of electric vehicles for long distance and commuting.

As a result of these developments, all major car manufacturers include at least one electric car model in their product line. The commercial availability of better-performing batteries for 2012 is expected to increase the size of the market and competition between electric vehicle manufacturers, which will result in lower prices for new vehicles and the production of conversion kits allowing for the conversion of many existing vehicles for they operate on electricity.

Australia currently consumes more than 20 billion litres of gasoline and 10 billion litres of diesel per year. Replacing even 10% of these volumes with Australian-produced biofuels would require the diversion of food crops for human consumption to fuel production to the point of distorting the domestic and export food markets. The availability of food crops would decrease, which would force the price.

As a result of these increases, farmers would use much more of their land to increase agricultural production, which would give them better yields. This would involve extracting production crops at a lower price. These, in turn, would become rare and increase prices until a new price equilibrium is found for agricultural food crops.

In the United States, this type of market distortion has caused a shortage and an increase in the cost of some foods, including sugar and cereals. This, in turn, has inflated the prices of meat, bakery products, and other food products that contain sugar and cereals.

In Australia, this problem has been avoided so far by importing biofuels, mainly from Indonesia and Malaysia. This has led to the destruction of the rainforest in these countries and the constant reduction of the habitat of flora and fauna, including the endangered orangutan. The cleared land is planted with crops such as oil palm, mainly for the production of biofuels. This leads to a rapid depletion of soil fertility and increased CO_2 emissions from the destruction of the rainforest, a short-term gain for long-term loss, which may be permanent.

There is a possibility of producing biofuels that would avoid these problems. That is, produce them from algae if they can be genetically modified to sequester the CO_2

present in the atmosphere and use it to produce a biofuel. However, it seems very unlikely that production of this means is as profitable as electricity.

Unless biofuels can compete with electricity, their production of propulsion is not an economic reason, which would only delay the return process of the internal combustion engine to the waste pile of history. However, the production of bio-oil can be valuable as a raw material for the manufacture of fertilizers, plastics, and other petrochemical products.

Some people claim that electricity is generated mainly by burning fossil fuels such as coal, gas or oil, the primary sources of CO_2 emissions. Therefore, an increase in the demand for electricity to power automobiles and other vehicles would increase these emissions. Biofuels would not be. This statement is valid only if the electricity used to recharge vehicle batteries is produced from fossil fuels. Currently, biofuels are not used for this purpose.

Australia's electricity needs are starting to be generated from renewable sources such as hydropower, wind power, solar power and the enormous potential of geothermal energy. The government has mandated that 20% of our electricity needs come from renewable sources by 2020. This goal can be exceeded well before 2020, thus recharging the battery without increasing CO_2 emissions. Also, most refills can be done at night; otherwise, a large part of the electricity produced would be wasted.

Many other countries, are in a similar situation in Australia. Some have higher renewable energy targets and others, like France, generate almost all of their electricity needs without using fossil fuels.

Although biofuels emit only CO_2 extracted from the atmosphere by the plants from which they come, they still produce emissions. Electricity produced from other renewable sources is cheap and more importantly.

The only reason to choose to produce or use biofuels for propulsion would be a strange desire to cling to the use of the internal combustion engine and the associated pollution.

This price is too high. We do not need biofuels. Electricity is a much better option for vehicle propulsion to be cheaper, widely available and much more efficient.

1.5 Downsides and benefits of using biodiesel

Biofuels differ from fossil fuels in several important ways.

• They are a type of total renewable energy.

• They emit much fewer greenhouse gases.

• They can occur in a growing season, unlike the fossil years that took millions of years to form.

There are different "generations" of biofuels.

• First generation biofuels come from sugar, other starches, and animal and vegetable oils. Examples include biodiesel and biogas.

• Second generation oils are derived from industrial waste, such as wood chips. Biofuel based on ethanol, other alcohols, and diesel engines belong to this category.

• The algae-based biofuel is the third generation. These are highly renewable because algae can quickly grow on a large scale and decompose quickly and easily.

• Microorganisms are used in fourth-generation biofuels. Like the third generation, they decompose rapidly and, therefore, have a low carbon footprint.

Now that we have learned the different types, let's look at the benefits of biofuels.

1.5.1 Benefits of Biodiesel

With the rising price of gasoline and our country's dependence on foreign fuels, are you looking for an alternative fuel source? Biodiesel can be the fuel source you are looking for. What biodiesel are you asking for? Biodiesel is a source of alternative fuel based on vegetable oil. You can use this fuel in any vehicle that uses regular diesel. There are places in some states where you can buy biodiesel fuel, but you can also do it yourself with a biodiesel kit.

This alternative fuel source has many advantages over oil. An advantage is a price. Biodiesel can be produced for much less than its petroleum equivalent. If you do it yourself, you should buy a biodiesel kit. You can get it for a few thousand dollars. You may think "Wow, it's a lot of money". But if you think about it, once the savings on gas prices add to the cost of the kit, all the savings go directly to your wallet. This will undoubtedly put some room for manoeuvre in your budget. If you do not make biodiesel yourself and you're lucky enough to live in an area where there is a biodiesel plant, your savings will accumulate much more quickly.

Another significant advantage of this alternative fuel is that it is very environmentally friendly. It is completely biodegradable. Emissions are almost non-existent. Biodiesel fuel burns cleanly and is not toxic. This means that it not only helps the planet, but does not harm humans. Since biodiesel does not contain sulphur either, its emissions do not contribute to acid rain. The use of biodiesel fuel also reduces cancer that causes the release of particles into the air.

One of the benefits closely related to the environmental issue is that biodiesel is a renewable resource. It means that it can be done over and over again. It does not have to be pumped out of the ground. Do you need more biodiesel fuel? Just plant another crop and you can do more. This presents a hidden advantage that allows farmers to make their farms profitable. It contributes to the economy and the environment.

Biodiesel has many advantages as an alternative source of fuel. So, is the use of biodiesel a disadvantage? The only thing to do is that if you have a vehicle manufactured before 1994, you have to watch the hoses of the fuel system closely. Biodiesel can soften the rubber used to make the pipes. After 1994, most vehicles were made with synthetic hoses and this was not a problem. Another problem that you should keep in mind is that the first few times you use biodiesel, you risk having a clogged food system. The biodiesel fuel will clean all the old dirt from your fuel system. It will remove any debris left on the walls of your fuel tank, which can obstruct some of the lines. If you are aware of it and are careful, you should not have any problems.

As an alternative source of fuel, biodiesel is first class. It is a renewable source because it is made of vegetable oil, it only cultivates another culture. This allows farmers to make their farms profitable (another hidden advantage). It's cheaper to produce, good for your wallet. It is environmentally friendly and does not cause a lot of problems such as petroleum fuels. It is also favourable to humans unless there are problems that cause cancer, as well as breathing problems. It is also safer to carry and store. It also reduces our country's reliance on foreign oil, which would help our economy.

1.5.2 Disadvantages of biofuels:

The creation of first-generation biofuels requires enormous amounts of land. Crops tend to be small varieties of the original, such as corn, which makes it unsuitable for human consumption. This reduces the area of cultivation to obtain good food crops and reduces the production of food. The price of edible corn has increased in recent years. This has been a hot topic of discussion with many arguments from both sides. Those who protest against the large-scale production of food crops for the production of biofuels are convinced that food should only be used for consumption. They argue that the production of biofuels has already had a negative impact on many poor people.

The creation of biofuels also requires vast amounts of equipment. With a large amount of land devoted to this, there is less left to grow edible food. Therefore, as more and more land is used for food crops and biofuels, there is less and less a natural habit of plant and animal ecosystems.

Another problem is that most biofuels are produced in relatively small quantities compared to large-scale production of coal, oil and natural gas. It is difficult to maintain a steady supply of biofuel, so you cannot always rely on it without a reliable source.

For these reasons, many believe that the best future use of biofuels will be in developing countries. Here, where fossil fuels can be scarce or unaffordable small-scale biofuel production can be a viable form of renewable energy. In the Indian state of Bihar, for example, smart citizens convert the biomass of human waste into biogas for the generation of electricity.

1.6 Sustainable of biodiesel

Biofuels are energy sources derived from organic materials that are more environmentally friendly and renewable. Unlike fossil fuels such as oil, coal, and gas, whose presence continues to shrink, biofuels have a future perspective of replacing fossil fuels as the primary source of energy. The most common biofuel is bioethanol. It is increasingly produced from sugar cane. To produce ethanol, the biofuel undergoes a fermentation process. Currently, ethanol provides less than 10% of consumption and will continue to increase by up to 30% by 2025 (source: International Energy Agency)

Currently, most researchers are not doing much to find the viability of biofuels to make them economical and practical. The goal is to ensure that biofuels have a higher return, such as fossil fuel gasoline. Efficiency is a measure of the amount of energy that can be used (output) for a particular purpose, extracted from the input energy.

According to research, ethanol derived from corn has only 20% efficiency, compared to 75% for gasoline. Biodiesel fuel was registered with an energy efficiency of 69%. The positive side of several studies is a cellulose product from the plant, which is 85% efficient. It's higher than nuclear energy.

Information on the oil futures market on the New York Stock Exchange shows that oil prices fluctuate dramatically. Analysts predict that the availability of biofuels will affect oil prices. While other analysis continues to predict that oil prices will continue to rise due to source depletion. The analysis again indicated that if the government were to provide subsidies to encourage the distribution of biofuels and to promote environmental biofuels, the demand for gasoline and diesel would decline. Therefore, it will be linked to the price of fossil fuels.

Several countries, such as Brazil, are actively involved in the development of biofuels. Brazil has become the world's largest production of ethanol derived from sugar cane and produces 1.5 billion litres of ethanol per year. The United States, the world's largest oil-consuming country, is the second largest producer of biofuels after Brazil. The EU's biodiesel production capacity now exceeds four million tons. 80% of EU biofuels derived from rapeseed oil, soybean oil and a marginal amount of palm oil account for 20%. The technology of a biofuel is essential. This is because the need for renewable energy is increasing. The biofuel is energy from natural resources. This will avoid the effects of global warming soon.

2 how does a diesel engine work?

Introduction

Diesel engines have been integral machines and widely produced for more than 100 years. The appearance of the automobile at the beginning of the 20th century is due to many different inventors and innovators. Perhaps the most critical contributor is Rudolph Diesel (1858-1913), the creator of the diesel engine.

Diesel first became interested in engines when he was young. In the 1980s, he was working on ideas for a new type of engine that would run faster and more efficiently than the steam and coal engines of the time.

The goal was to develop an engine that could compress the air faster and, therefore, move a vehicle faster. After the mixed success, the diesel model was patented and implemented in the success of the first cars and trucks.

These types of engines are set in motion when the fuel is injected into the air that is already compressed.

First, the air is taken at the engine. Then that air is compressed from inside the engine by a "piston". The piston then returns to its position at the top of the engine when fuel is injected into the machine, which causes combustion. This force moves the piston down.

As the engine fills with exhaust, the piston moves again to eject the exhaust. This process is repeated quickly, and the engine can only run until it is interrupted (due to lack of fuel, lack of air or because of the cessation of the movement of the piston).

The gasoline engine, which was developed after the death of Diesel, is very similar to the diesel engine, with only the order of the process changed.

Without these engines, there would be no automobile revolution in the 20th century. They would have arrived over time, but the development and timeline of the evolution of automobiles would be drastically different.

Some contemporary observers are wondering if society sees the gradual elimination of diesel engines. With the emergence of hybrid vehicles and the possibility of cars running on renewable energy and electricity, diesel engines will be joined by newer engine models for daily use.

These diesel models are still vital machines and are always in demand, despite the changing scope of the automotive design. People who own a diesel engine can sell exchange or buy engines and other auto parts, even online. For some cars that need repair, a diesel engine is a missing component. A new or updated diesel engine can also affect the way a car works.

2.1 HOW DO DIESEL ENGINES WORK?

The essential difference between a diesel engine and a gasoline engine is that, in a diesel engine, fuel is sprayed into the combustion chambers through the nozzles of the fuel injectors just when the air in each chamber has been placed under a pressure so high that it is so hot that it becomes inflamed.

Below is a step-by-step view of what happens when you start a vehicle that runs on diesel.

❖ **You turn the key in the ignition**

Then wait until the engine accumulates enough heat in the cylinders for a satisfactory start. (Most vehicles have a little light that says "Wait," but a sensual computer voice can do the same job on some vehicles.) When turning the key, a process begins in which the fuel is injected into the cylinders under such high pressure that it heats the air in the cylinders all by itself. The time it takes to warm things up has been drastically reduced, probably not more than 1.5 seconds in moderate weather.

Diesel fuel is less volatile than gasoline and is easier to start if the combustion chamber is preheated, so manufacturers originally installed small incandescent spark plugs that ran on the battery to preheat the air in the cylinders when the engine started. The best fuel management techniques and the highest injection pressures now create enough heat to eliminate fuel without glow plugs, but the plugs are still there to control emissions: the extra heat they provide helps burn the fuel more efficient way Some vehicles still have these cameras, others do not, but the results are still the same.

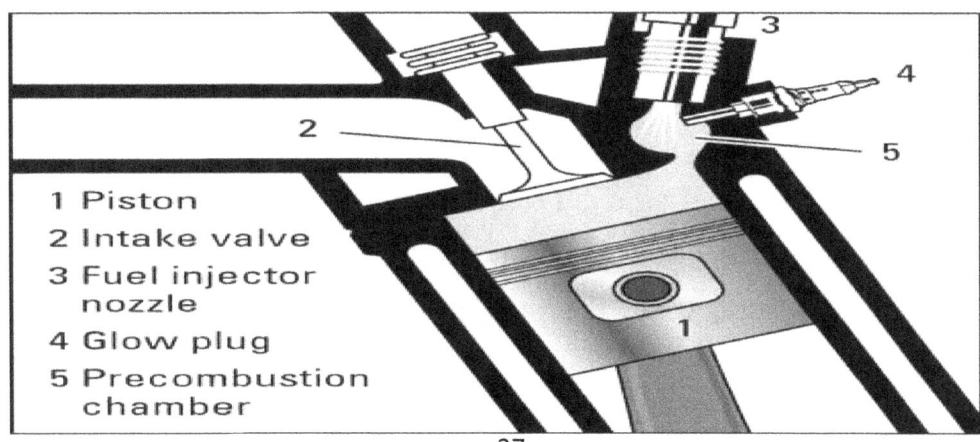

- ❖ **A "Start" light goes on**

When you see it, step on the accelerator and turn the ignition key to "Start".

- ❖ **Fuel pumps deliver the fuel from the fuel tank to the engine**

On its way, the fuel passes through a pair of fuel filters that clean it before it can reach the nozzles of the fuel injector. Proper maintenance of the filter is especially necessary for diesel engines since the fuel contamination can clog the small holes in the nozzles of the injectors.

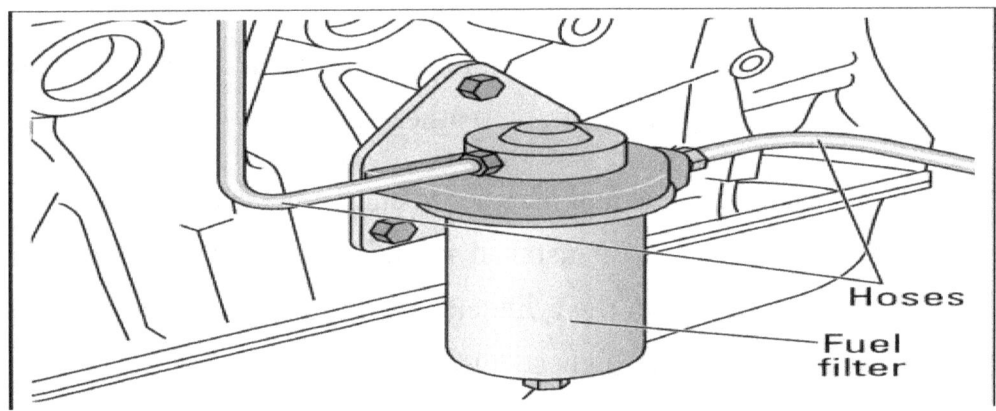

- ❖ **The fuel injection pump pressurizes fuel into a delivery tube**

This supply tube is called a rail and keeps it under a constant high pressure of 23,500 pounds per square inch (psi) or even higher while delivering the fuel to each cylinder at the right time. (Fuel injection pressure can be as low as 10 to 50 psi!) The fuel injectors feed the fuel as a fine mist into the combustion chambers of the cylinders through nozzles controlled by the engine control unit (ECU), which determines the pressure, when the fuel spraying occurs, its duration and other functions.

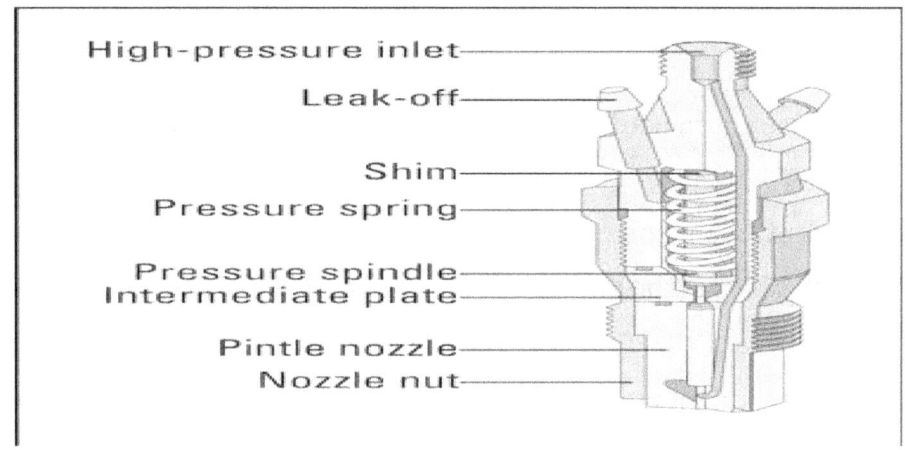

Other diesel fuel systems use hydraulic systems, crystalline wafers and other methods to control fuel injection, and others are being developed to produce diesel engines that are even more powerful and responsive.

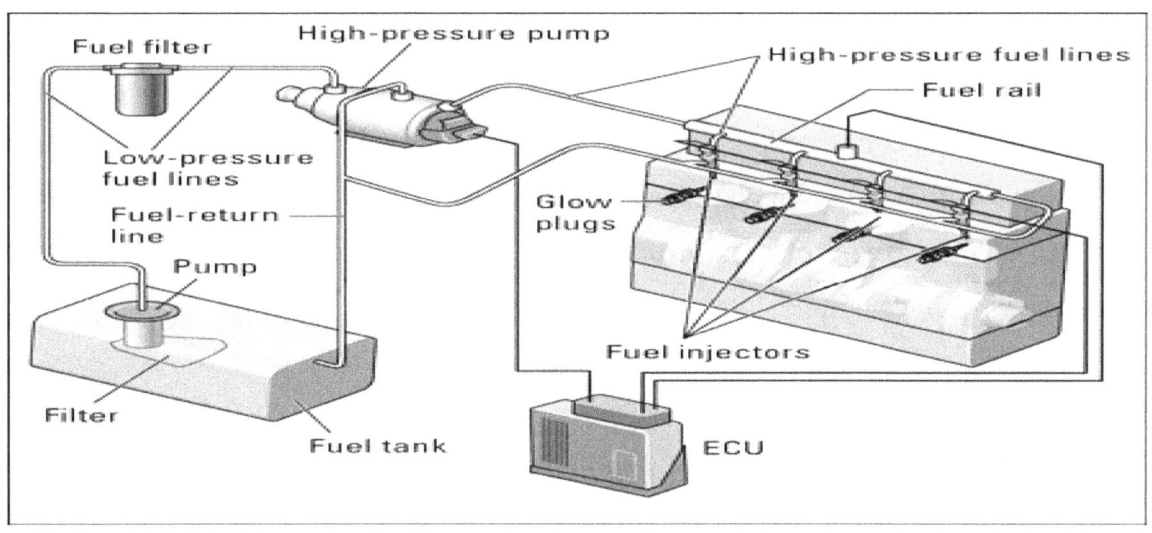

- ❖ **The fuel, air, and "fire" meet in the cylinders**

While the previous steps get the fuel where you need to go, another process runs simultaneously to get the air where you need to be in the last game of fiery power.

In conventional diesel engines, air enters through an air filter that is quite similar to that of gasoline vehicles. However, modern turbochargers can inject more significant volumes of air into the cylinders and can provide more power and fuel savings under optimal conditions. A turbocharger can increase the power of a diesel vehicle by 50 percent, while reducing its fuel consumption by 20 to 25 percent!

- ❖ The combustion propagates from the smallest amount of fuel pressurized in the pre-combustion chamber to the fuel and air in the combustion chamber itself.

2.2 DIESEL ENGINE

A diesel engine is an internal combustion engine, also known as a compression ignition engine. It uses the compression heat to start the ignition and then burns the fuel that is injected into the combustion chamber. Unlike the diesel engine, spark ignition engines use a spark plug to ignite a mixture of air and fuel. An example of a spark ignition engine is a gasoline engine.

The thermal efficiency of the diesel engine is the highest of any standard internal or external combustion engine. The highest thermal efficiency is the result of a high compression ratio.

Most reciprocating internal combustion engines operate in one of two mechanical cycles: the four-stroke cycle or the two-stroke cycle. These cycles designate, in the correct sequence, the mechanical actions by which fuel and air access the engine cylinder, the gas pressure, due to combustion, becomes power and, finally, the burned gas is expelled from the engine.

2.3 The basic four-stroke diesel engine

By its name, it is evident that there are four strokes in a complete engine cycle. A stroke is the movement of the piston along the entire length of the cylinder and, since one of these movements cause the crankshaft to turn half a turn, it follows that there are two revolutions of the crankshaft in a full engine cycle.

The four hits in the order in which they occur are:

1. Entrance race. With the inlet valve open and the exhaust valve closed, the piston moves from a top dead centre (TDC) to bottom dead centre (BDC), creating a low-pressure area in the cylinder. Clean, filtered air is precipitated through the open inlet valve to relieve this low-pressure area, and the cylinder fills with air.

2. Compression race. With both valves closed, the piston moves from BDC to TDC, compressing the air. During this race, the air is heated to a temperature high enough to ignite the fuel.

3. Blow of force. About TDC, the fuel is injected or sprayed into the compressed hot air, where it ignites, burns and expands. Both valves remain closed and the pressure acts on the piston crown, forcing it to lower the cylinder from TDC to BDC.

4. Escape race. At approximately BDC, the exhaust valve opens and the piston begins to move from BDC to TDC, expelling the burned gas from the cylinder through the open exhaust valve.

2.4 The two-stroke diesel

The 2-stroke engine uses two piston strokes to complete an influence stroke and, therefore, shoots double the maximum amount as an internal-combustion engine. A two-stroke engine is smaller and a lot of simple with fewer moving components. A two-stroke engine has the potential to provide double the ability of an internal-combustion engine of a similar size, however, thanks to the extra adjustment needed in an exceedingly two-stroke diesel, for instance, blowers and governors, they become costlier to provide. There has been a shift towards four-stroke diesel engines that became a lot of economic and smaller.

2.5 Types of Diesel Engine

Nowadays, many car manufacturers are opting for diesel engines. Different types of cars, from the smallest to the most luxurious, are now equipped with engines. And the best thing to do is that these engines are no longer considered "dirty". Diesel is now synonymous with the cleanest technologies regarding fuel and diesel not only takes great care of the engine but also boasts of high energy efficiency. Diesel was invented in 1893 by Rudolf Diesel and clean diesel technology was developed by Volkswagen and Mercedes-Benz, the German automakers.

However, if you are interested in diesel engines, you should also know the different types of identical engines. The first type of engine is a common rail direct injection. This is a relatively modern invention and is based on the principle of the direct injection system of an engine. IDRC will boost performance, take care of the engine and improve mileage. This was invented and developed by Robert Huber; the Fiat group used it extensively and was eventually sold to Robert Bosch, the German company. In the early 90's, it became extremely popular with city car buyers because of its many benefits.

The second type of diesel engine is the Multijet fuel injection. This too was developed by the Fiat group. And with this diesel engine system, IDRC's performance has been significantly improved. The engine uses an injector to receive fuel inputs into a combustion chamber. The injector is essentially a spring loaded valve that opens and closes at certain fuel pressures, thus producing energy. The multiple fuel injection points are synchronized and release a predetermined amount of fuel into the combustion chamber. Other types of engines are DI Turbocharged and Naturally Aspirated Engine.

2.6 Why You Should Consider Choosing a Diesel Engine Vehicle

While choosing the vehicle to buy, you can choose one with a diesel engine. Unlike the days when they were considered potent machines, diesel engines are more and more used and some motorists opt for them rather than the gasoline engine.

With the current technology, the defects of the old, smelly and noisy diesel engine have been reduced and it has become more user-friendly.As with gasoline-powered cars and SUVs, diesel-powered vehicles are fully equipped with modern safety features such as safety locks, front side airbags, full-length side curtain airbags, and stability.

Diesel engines were considered the most suitable for gross manual transmissions; they have now been modernized for use in the standard tramway. The transmissions have also been automated. They offer superior speed and improved fuel economy without losing the convenience of automatic touch.

Larger vehicles with larger passenger capacity like the SUV work better with the diesel engine because they are heavier. The devices are better suited to heavier tasks than gasoline engines.

The diesel engine consumes less fuel than the gasoline engine while performing the same task, due to the high combustion temperature and the higher expansion rate of the engine. They have a low voltage electrical ignition system, resulting in high reliability and easily adaptable to wet environments. The life of these engines is about twice as long as that of the gasoline engine because of the increased strength of the parts used. Diesel fuel has better lubricity properties than gasoline, which contributes to the excellent condition of the engine.

Diesel is also considered safer than gasoline. While gasoline creates flammable vapors in the open air and explodes easily, diesel fuel burns freely in the open with a wick and does not explode. Having this engine is, therefore, safer for your vehicle; he is less likely to explode if he is involved in an accident.

Diesel engines emit less heat during cooling and exhaust. The carbon monoxide content in the exhaust gas is minimal, which reduces the risk of carbon monoxide poisoning for the motorist and others who may inhale. As the combustion of the diesel engine is higher, there is more exhaust coming out of the exhaust. Obtaining the best exhaust systems for your vehicle will increase exhaust emissions from the engine and therefore improve performance.

There are suppliers like Magnaflow who specialize in diesel exhaust systems for vehicles such as the Dodge, Chevrolet, and Ford. Getting these engines is not only safe, but also helps to get the best performance from your vehicle.

2.7 Advantages of Diesel Engines

Diesel engines have certain advantages over gasoline engines, which make them more suitable for tasks that require a lot of power or torque. One of the main differences between a diesel engine and a gasoline engine is the way they start. In a diesel engine, fuel is pumped into the compression chamber after air compression. This causes the spontaneous ignition of the fuel, which avoids having to use spark plugs.

Also, these engines have larger pistons, which mean that combustion is more powerful. This leads to the need for stronger parts withstand the pressure, and the stronger parts generally mean more substantial parts. This is why diesel engines are not used in airplanes the weight is too much

In a gasoline engine, fuel and air are mixed in the intake manifold and sucked into the compression chamber. Then they need on by a spark. While gasoline engines may have more speed, especially when it comes to starting from a stationary position, they do not have the same power. For this reason, diesel engines are the ideal choice for towing caravans or boats or for driving larger and heavier vehicles, such as trucks or buses.

Diesel engines have fewer moving parts and, therefore, are unlikely to wear out at the same rate as other types of engines.

You will get better fuel economy with a diesel engine due to the higher diesel fuel density. At a time when fuel prices seem to be increasing daily, this is an important consideration. Not only uses less fuel, but the price of that fuel is cheaper, at least so far, so it is saved on two fronts. Many people do not realize that it is possible to modify the performance of the engine to do it faster, without damaging the fuel economy.

In the past, it was considered that the engines were worse to leave the contamination. But many manufacturers are now using a new technology to solve that problem and it is less likely that newer engines will displace much smoke. Another important feature that can be placed at the foot of the new technology is that you can now get better acceleration speeds in the newer diesel engines while maintaining the same good fuel economy.

In some countries, the pollution caused by diesel is due to the high sulphur content.This type of diesel is a very cheap grade, and refiners will take a while to replace it with the higher grade diesel that contains less sulphur. Until this happens, diesel will likely remain a secondary fuel option in those countries, especially when pollution problems are given a higher priority. In many European countries, diesel cars are much more common than in Western countries.

2.8 Diesel Engine Problems

The fundamental disadvantage of the diesel engine is that it is expensive. It is expensive both in manufacturing (due to the high workload) and in maintenance. It is expensive due to the ecological incompatibility of its escape and due to the need to adjust its escape by the strict requirements of international agreements.

The heat of the compressed air ignites the fuel in the diesel engine. The result is that the fuel did not have time to thoroughly mix with the air and produces CH, NOX and carbon black during the combustion process. Carbon black is particularly visible, so colour the exhaust in black.And if in the case of hydrocarbons can be removed with a catalyst, the amount of carbon black in the exhaust is adjusted by the special exhaust filter, which is mounted between the exhaust manifold and the catalyst. The exhaust filter is heating up in the exhaust gas stream which results in a combustion of carbon black. Periodically, the residual carbon black must burn and, according to the instructions of the command block, the temperature of the gas increases at the end of the combustion stroke due to the burning of an additional amount of fuel.

The catalyst has a more complex design due to the irregular chemistry of the exhaust gases.

Now let's look at the problems related to diesel fuel. As you know, diesel fuel is of 2 types: summer fuel and winter fuel. They differ in the solidification temperature. When the fuel freezes, the fuel pump cannot discharge it and that is the end ... You are idling at the edge of the road, unable to start the engine. This can be avoided by heating the fuel lines (also the truck fuel tank). Unlike diesel fuel, gasoline does not freeze.

The HPFP diesel engine is an extremely unreliable unit. Due to its operation at higher pressures, the entry of water into the fuel is a mortal danger. Therefore, water separator is required. Small particles of dirt can also damage the pump, so the filter is necessary after filling. In the Russian environment, 2 filters are required due to filthy diesel fuel in Russia and not all resellers provide cars equipped with such engines to Russia. This complexity of the engine systems results in higher prices for diesel engines; sometimes the price difference compared to gasoline engines is up to 4000 euros.

Noise and vibration until recent times could not be separated from the words "diesel engine". Attempts to neutralize them involve wrapping the engine compartment with acoustic insulation, balancing the engine moments and calibrating the control units.

2.9 Why diesel engines are still popular?

The first, ecological regulations are maintained in foreign countries and owners of eco-friendly cars have discounts on insurance and other taxes.

Secondly, under conditions of fuel quality and fuel maintenance in the regular base diesel engine, it can operate up to half a million kilometres without capital repair. And that is the sure gain.

The third, the turbo-supercharger diesel engine can undoubtedly play the role of "fire starter". Many car manufacturers follow that path.

2.10 Benefits of Diesel Engines

You can see diesel engines used mainly in commercial applications on large platforms or commercial vessels; however, much smaller and different cars and boats use this type of engine as well. This observation makes many people wonder why the two different options in the engine: standard gasoline and engine area. Diesel engines offer many different advantages over standard gasoline engines, but they also have many drawbacks, depending on how the engine is used. Trying to decide what work best for a particular circumstance willinvolve being fully aware of the benefits of these types of engines.

Compared to standard gasoline engines, diesel engines are much more efficient. They burn fuel with a much higher efficiency, which leads to a reduction in fuel consumption and a higher value for the dollars spent on fuel. This also includes heat insulation in the engine, which is significantly less than a standard gasoline engine. This improves the operation of the engine. This leads to greater efficiency in the delivery of power; Diesel engines have almost constant force delivery, which means more power on demand.

Due to their higher efficiency and construction, they also have a useful life that is almost double that of a standard engine. Also, the exhaust of these engines has a lower carbon emission rating. Therefore, these engines have a lower impact on the environment and the costs of a person or company.

These engines also have the advantage of being much more straightforward than standard gasoline engines. Simplicity means that there is less chance that something goes wrong and the engine breaks down. For these types, there is no electric ignition system, like the spark plugs, because the fuel is ignited only through the pressure of the pistons. This is a less system to decompose, but also makes the engine more resistant to breakdowns in humid environments. This factor makes a diesel engine perfect for commercial vehicles and boats.

Finally, these engines are also much less dangerous. Most people, just driving their cars or personal trucks, do not have to worry about fuel safety; however, for more extensive commercial operations that store and distribute their fuel, safety in this area is paramount. Diesel fuel is much less likely to ignite compared to gasoline fuel, so storage and handling are a much safer prospect.

Although it has many benefits, diesel engines have some drawbacks. That is, these engines are difficult to start in cold climates. Depending on the environment, these pros and cons can be carefully weighed to make the right purchase decision for you or your business.

3 Materials in the production of biodiesel

The raw material used in production of biodiesel

The raw materials for biodiesel production are vegetable oils, animal fats and short chain alcohols. The oils most used for worldwide biodiesel production are replaced (mainly in the European Union countries), soybean (Argentina and the United States of America), palm (Asian and Central American countries) and sunflower, although other oils are also used, including peanut, linseed, safflower, used vegetable oils, and also animal fats. Methanol is the most frequently used alcohol although, ethanol can also be used. Since cost is the main concern in biodiesel production and trading (mainly due to oil prices), the use of non-edible vegetable oils has been studied for several years with good results.

3.1 Introduction to Biodiesel Production

Besides its lower cost, another undeniable advantage of non-edible oils for biodiesel production lies in the fact that no foodstuffs are spent to produce fuel. These and other reasons have led to medium- and large-scale biodiesel production trials in several countries, using non-edible oils such as castor oil, Tung, cotton, jojoba and jatropha. Animal fats are also an interesting option, especially in countries with plenty of livestock resources, although it is necessary to carry out preliminary treatment since they are solid; furthermore, highly acidic grease from cattle, pork, poultry, and fish can be used. Microalgae appear to be a very important alternative for future biodiesel production due to their very high oil yield; however, it must be taken into account that only some species are useful for biofuel production. Although the properties of oils and fats used as raw materials may differ, the properties of biodiesel must be the same, complying with the requirements set by international standards.

3.2 Edible vegetable oils

Biodiesel has been produced primarily (over 95%) from edible vegetable oils (first-generation biodiesel) around the world, readily available on a large scale in the agricultural sector. Currently, biodiesel is primarily made from raised in Canada, soybean in the United States, sunflower in Europe and palm in Southeast Asia.

However, continued large-scale production of biodiesel from edible oils has recently raised serious concerns as they compete with food products. Knowing this, nearly 60% of humans in the world suffer from malnutrition.

The main biodiesel producers were the European Union, the United States, Brazil, and Indonesia, with a combined use of edible oil for biodiesel production of about 8.6 million tonnes (7.8 million hectares) in 2007. The estimated increase in edible oil used for biodiesel production was 6.6 million tonnes between 2004 and 2007, which would account for 34% of the increase in consumption.

Between 2005 and 2017, the use of edible oils into biodiesel is expected to account for more than a third of the expected growth in the use of edible oils, which means an increase in the price of biomass, an increase in water problem of water availability, and in particular a more extensive area available. Somewhere in the world will be converted to farmland, thereby releasing GHG emissions. Because of these drawbacks, the researchers looked for other renewable resources for biodiesel production.

3.3 Inedible vegetable oils

Technologies are being developed to exploit cellulosic materials used in the production of biodiesel (biodiesel, second generation), such as leaves and stems of plants, biomass from waste, as well as oilseeds from non-biodiesel plants. Non-edible biodiesel crops are expected to use mostly unproductive land and those located in poverty are sanded from degraded forests. Also, non-edible oilseeds are well adapted to arid and semi-arid conditions and require low fertility and a demand for moisture to grow. Also, inedible oils are not suitable for human consumption because of the presence of toxic components in the oils. For all these reasons, the use of non-edible oils as a raw material is a promising way to produce biodiesel.

There are a large number of oil plants producing inedible oils. On a list of 75 plant species containing oil in their seeds or nuclei, 26 species were reported by Azam et al as potential sources of biodiesel production. The central non-edible oleaginous plants are jatropha, Karanja, tobacco, Mahua, neem, rubber, sea mango, beaver, and cotton. Among these raw materials, jatropha, Moringa and castor oils are the most used in the production of biodiesel?

In Algeria, castor oil is recoverable, but it is not very common. Besides, the biodiesel produced from castor oil has a very high viscosity value compared to the value imposed by the American standard (ASTM D6751) and the European standard (EN14214). The meringue was planted in Mascara (359 km west of Algiers), where the climate of the region was not conducive to its development. Also, Moringa has just been planted in Tamenrasset (1970 km south of Algiers) and the Technical Institute of Fruit Growing ITAF (Algiers), but its development potential is not yet put in evidence.. Jatropha was planted in Adrar (1543 km south of Algiers) as part of the JatroMed project, which aims to cultivate jatropha to control its development potential in Algeria. JatroMed involves five countries in the Mediterranean region: Greece (project coordinator), Italy, Egypt, Morocco, and Algeria.

Inedible oil plants are called upon to solve the problem of competition with food production. However, the problem of water needs, water availability and mainly the amount of GHG generated by the high rate of exploitable land could not be solved with this raw material.

3.4 Used Edible Oils

There are several end uses of used edible oils (commonly referred to as used cooking oils), such as the production of soaps or energy by anaerobic digestion or thermal cracking. However, due to the poor quality of the soap produced by the WCO, a large number of clients have been found to be 338 A.C. Amir et al. Large quantities of the MDGs are illegally dumped into rivers and landfills, resulting in environmental pollution, Therefore, the management of these oils and fats poses a significant challenge due to their disposal problems and possible contamination water and land resources.

The WCO's biodiesel production to partially replace petroleum diesel is one of the measures taken to address the dual problem of environmental pollution and energy scarcity. Also, to reduce biodiesel production costs, the WCO would be a good choice as a raw material because it is cheaper than virgin vegetable oils and other raw materials.

The edible oil used is classified according to its free fatty acid (FFA) content. If the free fatty acid content of the OMD is <15%, then it is called "yellow fat"; otherwise, it is called "brown fat". The amount of MDGs generated in each country is enormous and varies depending on the use of vegetable oil.

3.5 Microalgae

Microalgae as raw material for biodiesel (third-generation biodiesel) have been the subject of a thorough review in recent years. They are photosynthetic microorganisms that convert sunlight, water, and CO_2 into algal biomass. Microalgae are classified as diatoms (Bacillariophyceae), green algae (Chlorophyceae), brown algae (Chrysophyceae) and blue-green algae (Cyanophyceae).

Microalgae have long been recognized as potentially a good source of biofuel production because of their high oil content (over 20%) and their rapid production of biomass. Algae biomass can play an essential role in solving the problem between food production and biofuel production soon. Microalgae cultivation does not need a lot of soil compared to terrane plants.

Due to their high viscosity (about 10 to 20 times higher than that of diesel fuel) and their low volatility, microalgae do not burn completely and form deposits in the fuel injector of diesel engines. Tran's esterification of microalgae oils will significantly reduce initial viscosity and increase fluidity.

3.6 Animal Fats

Animal fats used to produce biodiesel include tallow, white fat or lard of choice, fish fat (in Japan) and chicken fat. Compared with cultivated plants, these fats often offer an economic advantage because their cost price for conversion to biodiesel is favourable. The methyl ester of animal fat has certain advantages such as a high certain number, non-corrosive, clean and renewable properties. Animal fats tend to be low in FFA and water, but their amount is limited, which means they could never meet the world's fuel needs.

3.7 Materials

3.7.1 Materials handling: Oil

- Avoid using open drums for transport
- Utilize spill containment
- Use oil-dry or sawdust for clean up
- Avoid washing into storm drains or ditches

3.7.2 Materials handling: Methanol

- Overexposure to methanol can cause neurological damage and other health problems; and it presents a serious fire risk
- 1,000 ppm will produce symptoms such as irritation of the eyes and mucous membranes
- 5,000 ppm will result in a stupor or sleepiness
- 50,000 ppm will result in narcosis (deep unconsciousness) in one or two hours, probably resulting in death
- Handling methanol is the biggest fire threat
- Static can build up in PVC or plastic containers
- Pumps, motors should be sealed

3.7.3 Materials handling: Catalyst

- Sodium hydroxide (NaOH) and potassium hydroxide (KOH) are corrosive and may be fatal if ingested
- Skin contact can cause severe burns and should be thoroughly flushed with water or a dilute vinegar solution
- Inhalation of the solid NaOH or KOH is possible in dust-sized particles
- NaOH, KOH and concentrated solutions should never contact aluminuim, as they will create explosive hydrogen gas

- Fine particles of NaOH or KOH will produce holes in clothing, thus a protective apron or jumpsuit is advised

3.7.4 Waste stream handling

- Glycerol: Have a disposal plan burns or use as fertilizer
- DO NOT DUMP: Wash Water
- Keep out of the septic system: Trash, rags
- Can combust spontaneously

3.8 How to transport the material

As with petroleum, the B100 must be transported in a way that does not cause contamination. The following procedures are recommended for trucks and wagons and are used by biodiesel based distributors and transporters:

- Make sure the trucks and wagons are made of aluminium, carbon steel or stainless steel.
- Make sure you perform an inspection or washing (washing certificate) before loading.
- Check for previous and residual loading. In general, only diesel fuel or biodiesel is acceptable as waste. If the container has not been washed, some residues (including food, raw vegetable oils, gasoline or lubricants) may not be acceptable.
- Make sure there is no residual water in the tank.
- Check that the hoses and seals are clean and that they are made of materials compatible with the B100.
- Determine the need for insulation or a method of heating the contents of trucks or wagons if this is done in cold climates.

The B100 is a challenge to send in cold weather. In winter, most B100 models are shipped in one of the following ways:

- Warm (or at least hot) in trucks for immediate delivery from 27° to 54°C (80° to 130°F)
- Warm, around 50°C (122°F), in wagons for delivery in seven to eight days (it gets very hot if it has passed a single week since loading)
- Frozen in wagons equipped with external steam coils (tank wagon fuel melts to the final steam destination)
- In a mixture with winter diesel, kerosene or other low filling point in wagons or trucks.

No matter how biodiesel reaches, procedures should be used to prevent the B100 temperature drop below its obscurity for storage and handling. The biodiesel cloud point, ambient temperature and transport time should be taken into account when transporting the B100 to ensure that the fuel does not freeze.

3.9 Health and safety

> ➢ Always wear safety goggles and gloves when handling catalyst, or when mixing your catalyst and alcohol. Never rely on your reading glasses for eye protection; use goggles that fully seal and protect your eyes from potential splashes.

- Keep upwind of vented containers used to mix your catalyst with your alcohol, and do not work in an area that has an open flame source such as a gas water heater.

- Note that there are gloves rated to withstand breakthrough of methanol and sodium hydroxide; make sure the gloves you use will address your needs. Also, understand that glove will deteriorate with use and exposure over time.

- NEVER look directly into a container opening that you are using to mix chemicals. Chemical reactions could potentially cause chemicals to splash and get onto your skin and face; this compound can blind you if it gets in your eyes.

- Never dump copious amounts of sodium (or potassium) hydroxide to your alcohol all at once. Add the catalyst slowly, and consider mixing your solution intermittently during the process. This will prevent the "volcano" effect which may occur as a result of the heat released by the reaction of your catalyst with the methanol. If not controlled or if added to methanol that is already warm, this heat of the reaction can cause the methanol to boil over, or even catch fire. Obviously, this poses a personal protection, risk to the chemist mixing the ingredients, and also results in wasted materials. I strongly recommend you have running water nearby in the event you get sodium or potassium methoxide in your eyes or on your skin.

- You are likely to use braided tubing in your processor as a sight tube. Keep an eye on this tubing for bubbling. This will result when biodiesel gets between the layers of tubing that are bound together by the braiding, and is a sure indication that you will soon have a leak. If you see any bubbling, replace the tubing as soon as possible.. To avoid bubbling, drain fluids from sections of your processor while you are waiting for the transesterification to complete.

- If possible, replace the braided tubing in your processor with black iron pipe. In fact, I'd recommend replacing any component of your processing that comes into contact with biodiesel with black iron since this will eliminate weak points in your system. Depending on how you are set up, you can always measure the amount of vegetable oil you are adding to your processor independently of a sight tube.

➢ Avoid using PVC tubing with any part of your setup that comes into contact with biodiesel or vegetable oil. In my experience, biodiesel will cause it to become brittle and crack over time. Needless to say, this can lead to a big mess.

➢ Make sure you have some kind of spill containment system for vegetable oil, glycerin, or biodiesel. This can be as simple as building a soil berm around your processor, or as sophisticated as purchasing a commercial spill containment system such as the ones available through US Plastics.

➢ If you do have a spill, you'll want to be prepared to recuperate the media as soon as possible. A dust pan works well for shallow spills, and you'll want some empty buckets nearby ready for collection. If your finished biodiesel gets spilled on dry ground, you should be able to collect, filter and use it with no issues. Always have a dustpan and bucket/barrel nearby so that you can contain any spills as soon as they occur.

- Keep biodiesel off any concrete footers or floors. I've noticed concrete will crumble after repeated exposure to biodiesel. I'm assuming this applies to your home foundation or garage floor as well.

- Always have fire extinguishers nearby. If you have a fire, it will likely be either a Class B or Class C fire. A Class B fire results from flammable or combustible liquids such as gasoline, kerosene, grease and oil. Class C fires involve electrical

-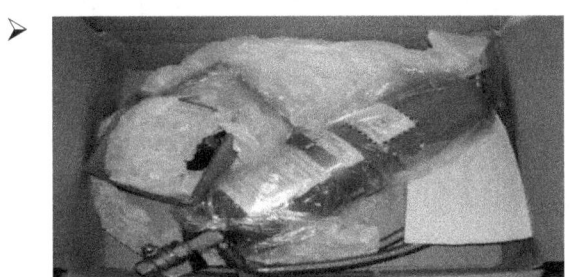

- equipment, such as appliances, wiring, circuit breakers and outlets. Given the risk of electrical shock, never use water to extinguish class C fires.

- Fire extinguishers containing dry chemicals are rated to handle class BC fires. These are typically filled with sodium bicarbonate or potassium bicarbonate. Note that the BC variety of extinguisher leaves a mildly corrosive residue which must be cleaned immediately to prevent any damage to materials.

- If you are trying to avoid scrutiny from your neighbours, DO NOT move your operation into a confined area shared by a gas furnace or water heater; methanol fumes associated with your processing are very likely to find the flame source. If you don't have a flame source in your garage or shed and you decide you must pursue this indoors, make sure you have adequate ventilation and positive pressure in your work area (i.e. An open garage door with a fan operating on high), and that combustible material such as oily rags are disposed of appropriately.

- Make sure that the pump you use with your processor has a brass impeller. This will minimize risk associated with sparking which can trigger combustion of alcohol.

> Note that although Harbor Freight pumps contain brass impellers, these are NOT "UL" rated for the purpose of making biodiesel. This means that they are not certified by Underwriters Laboratory (UL) - an independent, not-for-profit product safety testing and certification organization. Although I personally have had success using these pumps, I do wish to make this distinction clear

> Do not store copious amounts of methanol or ethanol in a residential space. If you must store alcohol, store no more than 10 gallons at any given time in appropriate containers rated for handling fuel.

If you are asked by the fire department what flammable liquids you are using and how they are classified, consider providing them with the following information:

Methanol – Class IB Flammable Liquid Sodium

Methylate – Class IB Flammable Liquid

Vegetable Oils – Class IIIB Combustible Liquid

Yellow Grease (Used Cooking Oils) – Class IIIB Combustible Liquid

Biodiesel – Class IIIB Combustible Liquid

Glycerin – Class IIIB Combustible Liquid

Sulfuric Acid – Class IIIB

3.9.1 Definitions

- Flash Point: The minimum temperature at which a liquid gives off vapor in sufficient concentration to form an ignitable mixture with air near the surface of the liquid (under controlled test conditions).

- Flammable Liquids: Defined as liquids having closed cup flash points below 100°F (37°C) and vapor pressures not exceeding 40 psi (276 kPa) (2.76 bar) at 100°F (37°C). Flammable liquids are referred to as Class 1 liquids.

Class IA liquids - flash points below 73°F (22.8°C) and boiling points below 100°F (37.8°C).

Class IB liquids - flash points below 73°F (22.8°C) and boiling points at or above 100°F (37.8°C).

Class IC liquids - flash points at or above 73°F (22.8°C) and below 100°F (37.8°C).

- Combustible Liquids: Defined as liquids having closed cup flash points at or above 100°F (37°C). Combustible liquids are referred to as Class II or Class III liquids.

Class II liquids - flash points at or above 100°F (37.8°C) and below 140°F (60°C).

Class IIIA liquids - flash points at or above 140°F (60°C) and below 200°F (93.4°C).

Class IIIB liquids - flash points at or above 200°F (93.4°C).

3.9.2 Fire Extinguishers

Fire extinguishers are divided into four categories, based on different types of fires. Each fire extinguisher also has a numerical rating that serves as a guide for the amount of fire the extinguisher can handle.

Class A extinguishers are for ordinary combustible materials such as paper, wood, cardboard, and most plastics. The numerical rating on these types of extinguishers indicates the amount of water it holds and the amount of fire it can extinguish.

Class B fires involve flammable or combustible liquids such as gasoline, kerosene, grease and oil. The numerical rating for class B extinguishers indicates the approximate number of square feet of fire it can extinguish.

Class C fires involve electrical equipment, such as appliances, wiring, circuit breakers and outlets. Never use water to extinguish class C fires - the risk of electrical shock is far too great!

Class C extinguishers do not have a numerical rating. The C classification means the extinguishing agent is non-conductive.

Class D fire extinguishers are commonly found in a chemical laboratory. They are for fires that involve combustible metals, such as magnesium, titanium, potassium and sodium. These types of extinguishers also have no numerical rating, nor are they given a multi-purpose rating - they are designed for class D fires only.

Water extinguishers or APW extinguishers (air-pressurized water) are suitable for class A fires only. Try to avoid using a water extinguisher on grease fires, electrical fires or class D fires – the flames will spread and make the fire bigger! (Personal note: I have discussed this with firefighters and they note that if all you have is water available, consider using it to extinguish the fire. I'm informed that the way to do this is to migrate the fire towards an obstruction where it can be cornered and doused. Having said so, I take no responsibility for any fire and/or method you use to extinguish a fire you may encounter).

Dry chemical extinguishers come in a variety of types and are suitable for a combination of class A, B and C fires. These are filled with foam or powder and pressurized with nitrogen. Dry chemical extinguishers have an advantage over CO2 extinguishers since they leave a nonflammable substantial on the extinguished material, reducing the likelihood of re-ignition.

BC - This is the regular type of dry chemical extinguisher. It is filled with sodium bicarbonate or potassium bicarbonate. The BC variety leaves a mildly corrosive residue which must be cleaned immediately to prevent any damage to materials.

ABC - This is the multipurpose dry chemical extinguisher. The ABC type is filled with mono ammonium phosphate, a yellow powder that leaves a sticky residue that may be damaging to electrical appliances such as a computer.

Carbon Dioxide (CO2) extinguishers are used for class B and C fires. CO2 extinguishers contain carbon dioxide, a non-flammable gas, and are highly pressurized. CO2 extinguishers displace oxygen feeding the fire. The pressure is so great that it is not uncommon for bits of dry ice to shoot out the nozzle. They don't work very well on class A fires because they may not be able to displace enough oxygen to put the fire out.

Using the fire extinguisher: Pull the Pin at the top of the extinguisher. The pin releases a locking mechanism and will allow you to discharge the extinguisher. Aim at the base of the fire, not them flames. This is important - in order to put out the fire, you must extinguish the fuel. Squeeze the lever slowly. This will release the extinguishing agent in the extinguisher. If the handle is released, the discharge will stop. Sweep from side to side. Using a sweeping motion, move the fire extinguisher back and forth until the fire is completely out. Operate the extinguisher from a safe distance, several feet away, and then move towards the fire once it starts to diminish. Be sure to read the instructions on your fire extinguisher - different fire extinguishers recommend operating them from different distances.

Remember: Aim at the base of the fire, not at the flames!!!!

3.10 Considerations Regarding Methanol

Methanol is the most dangerous chemical we use in making biodiesel and extreme care should be used in its handling and storage. Methanol drums should be grounded. There are grounding cables with big heavy clips that would work well for this. Don't ground to plumbing or conduit. The proper way to ground is to a copper stake driven into the ground next to the drum. If draining or pumping into a metal container the drum should be bonded (electrically connected) to the container.

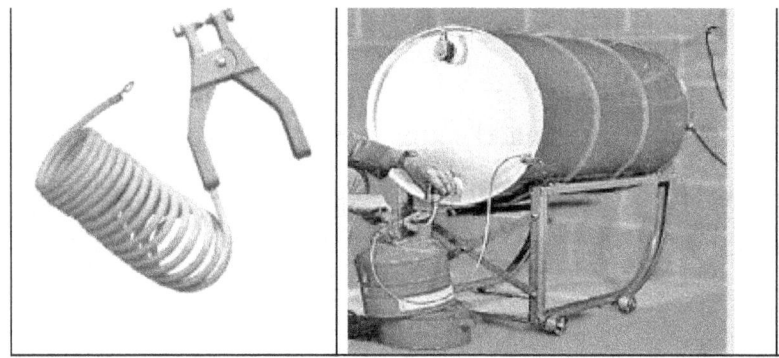

Methanol should be stored away from everything in a cool shady place.

The drum should have a secondary containment or drip pan to catch spills. Spillcontainment canbe a large plastic bucket placed under the drum. There are "over packs" available that perform thetask of spill containment. There any number of commercially available containments that willcatch a spill. More economical are plastic tubs or trays from the farm store used for wateringlivestock. If the drum is on its side, with a drain, then it will need a drip pan to catch any dripsout of the spigot.

Empty methanol drums are more likely to explode than full drums. That's because there is verylittle thermal mass to absorb the heat. An empty drum when exposed to the intense heat of a firewill quickly heat up to 386°F (the auto ignition temperature of methanol). When it does themethanol vapors inside the drum ignite and rip the drum apart at the seams. The lid is usuallylaunched high into the air and will cut through plywood roofs like they were butter. As such,return them right away for your drum deposit. If you must store the drums for a short time,remove the bungs, rinse them out with water, and store them upside down, without the bungs, soit drains. The same can be done with five gallon carboys used to store smaller amounts ofmethanol.

3.10.1 EMERGENCY OVERVIEW: Appearance: clear, colorless.

Flash Point: 11 deg C. Poison cannot be made non-poisonous. It causes eye and skin irritation. It may be absorbed through intact skin. This substance has caused adverse reproductive and fatal effects in animals.

Flammable liquid and vapor: Harmful if inhaled. May be fatal or cause blindness if swallowed. It may cause centralnervous system depression. It may also cause digestive tract irritation with nausea, vomiting, and diarrhoea. It causes respiratory tract irritation. It may cause liver, kidney and heart damage.

Target Organs: such as Kidneys, heart, central nervous system, liver, eyes. Potential Health Effects

Eye: Producesirritation, characterized by a burning sensation, redness, tearing, inflammation, and possible cornealinjury. It may cause painful sensitivity to light.

Skin: Causes moderate skin irritation. It may beabsorbed through the skin in harmful amounts. Prolonged and/or repeated contact may causedefatting of the skin and dermatitis. Ingestion: May be fatal or cause blindness if swallowed. It maycause gastrointestinal irritation with nausea, vomiting and diarrhoea. It may cause systemic toxicity withacidosis. It also causes central nervous system depression, characterized by excitement, followed byheadache, dizziness, drowsiness, and nausea. Advanced stages may cause collapse, unconsciousness,coma and possible death due to respiratory failure. It causes cardiopulmonary system effects.

Inhalation: Harmful if inhaled. It causes adverse central nervous system effects, including headache,convulsions, and possible death. Also causes visual impairment and possible permanent blindness.It causes irritation of the mucous membrane.

Chronic: Prolonged or repeated skin contact may causedermatitis. Chronic inhalation and ingestion may cause effects similar to those of acute inhalation andingestion. Chronic exposure may cause reproductive disorders and teratogenic effects. Laboratoryexperiments have resulted in mutagenic effects. Prolonged exposure may cause liver, kidney, andheart damage.

3.10.2 FIRST AID MEASURES: Eyes: Immediately flush eyes with plenty of water for at least 15 minutes,occasionally lifting the upper and lower eyelids. Get medical aid immediately. Skin: Immediately flushskin with plenty of soap and water for at least 15 minutes while removing contaminated clothing andshoes. Get medical aid if irritation develops or persists. Wash clothing before reuse. Ingestion: Ifvictim is conscious and alert, give 2-4 cupful of milk or water. Never give anything by mouth to anunconscious person. Get medical aid immediately. Induce vomiting by giving one teaspoon of Syrup ofIpecac.

Inhalation: Get medical aid immediately. Remove from exposure to fresh air immediately. Ifbreathing is difficult, give oxygen. Do NOT use mouth-to-mouth resuscitation. If breathing has ceasedapply artificial respiration using oxygen and a suitable mechanical device such as a bag and a mask.

Notes to Physician: Effects may be delayed. Ethanol may inhibit methanol metabolism.

3.11 Physical and Chemical Properties of methanol

Physical State: Liquid

Appearance: clear, colourless

Odour: alcohol-like - weak odour

PH: Not available.

Vapor Pressure: 128 mm Hg @ 20 deg C

Vapor Density: 1.11 (Air=1)

Evaporation Rate: 5.2 (Ether=1)

Viscosity: 0.55 cP 20 deg C

Boiling Point: 64.7 deg C @ 760.00mm Hg

Freezing/Melting Point:-98 deg C

Auto ignition Temperature: 464 deg C (867.20 deg F)

Flash Point: 11 deg C (51.80 deg F)

Decomposition Temperature: Not available.

NFPA Rating: (estimated) Health: 1; Flammability: 3; Reactivity: 0

Explosion Limits, Lower: 6.0 vol %

Upper: 36.00 vol %

Solubility: miscible

Specific Gravity/Density:.7910g/cm3

Molecular Formula: CH_4O

Molecular Weight: 32.04

3.12 The Dangers of Oily Rags

The following is sourced to the Wyoming Department of Employment:

Oily rags left in a closed container can become a safety nightmare. Many people do not believe it can happen. For no apparent reason, fire erupts, usually during off hours – as the materials have had time to react. It is called spontaneous ignition, and preventing is part of your job.

Spontaneous ignition occurs when a combustible object is heated to its ignition temperature of a chemical reaction involving the oxygen in the air around us. This "oxidation "process creates heat that, if not dissipated, will build up until ignition occurs. Generally, this can happen when the materials are left in piles and the heat being generated in the pile cannot be released into the air.

A number of materials are moderately or highly subject to spontaneous heating and subsequent ignition. Some of those you may find in your work area include oil based paint in contact with rags, cotton, or other fibrous combustible material; rags that are damp with any one of a number of different types of oils, including vegetable oils; oily uniforms or work clothes, and paint scrapings, possible coming from a paint spray booth cleaning project.

The possibility of spontaneous ignition is great if the surrounding air is also warm and dry. The added heat, say from nearby machinery, or a non-insulated steam line, can either pre-heat the material, which in turn sets off the reaction, or can hasten ignition by adding even more heat to the combustible.

It is simple to prevent spontaneous ignition, since oxygen is needed for it to occur. Materials subject to spontaneous ignition should be stored in covered metal containers such as a rag safety can or trashcan. Admittedly the container will contain oxygen at first. However, the oxidation process will use up the reaction and the reaction will stop fire prevented.

Another strategy is to spread the combustible material out so the resulting heat can be dissipated rather than allowed to build up-again, fire prevented.

Proper housekeeping is the key to preventing fires. Remove debris from the building or vessel. Properly store combustibles in covered containers. Be sure the lids of containers remain in place they are there for a purpose. Fire not only damages property, it threatens lives. Furthermore, a business destroyed by fire does not need employees. By working to prevent fires, you also work to protect your job.

On a personal note, I have personally witnessed a fire resulting from an old rag that caught fire. The evidence is shown in the picture below:

The fire took place in the outdoors in a well-ventilated area, and demonstrates that you can NEVER be too conservative when it comes to proper disposal of oily rags. I hope sharing this photo with the broader community drives the point home. Don't let this happen to you – read the MSDS for biodiesel included with this manual and note the comments associated with oily rags. Of all the risks, I've found that this one to be the most serious for me personally.

3.13 Hazardous Material Classifications and Spills

In the event you have a spill of any materials related to your operation, it is important to review the Material Safety Data Sheets (MSDS) provided with your chemicals so that you can protect yourself during cleanup. I've provided some copies of MSDS in the appendix of this manual. Regarding the materials themselves, please review the hazard classification summaries from the MSDS. You'll notice that spills are not reportable to hazardous waste for the chemicals involved in biodiesel manufacture or for biodiesel itself unless you need help with cleanup. Get these cleaned up as soon as possible

3.14 Hazards Involving Biodiesel Production

In a non-sensitive Situational Awareness Bulletin published last month, the Michigan Intelligence Operations Center (MIOC) discussed the hazards involving biodiesel production. For the protection and safety of all emergency responders nationwide, the Emergency Management and Response—Information Sharing and Analysis Center (EMR-ISAC) has excerpted the following information from the MIOC bulletin. The recent rise in petroleum prices caused an increased interest in alternative fuels. Biodiesel is used increasingly as a diesel replacement because it can be manufactured from readily available ingredients such as vegetable oil, animal fat, or recycled restaurant cooking oil. The production of biodiesel does not require a great amount of space, and the process is not easily detectable outside of the process area. There have been only a small number of casualties reported nationwide as a result of biodiesel production. The overall process is legal and relatively safe when properly performed. The end product of biodiesel has hazards similar to regular diesel with byproducts that can pose harm to humans and animals if not correctly stored or disposed. If the processors are not careful, they can poison or burn themselves, and modifications to pressure vessels by inexperienced people can result in possible explosions. Historically, the most common threat to homemade biodiesel labs is the improper storage and disposal of byproducts. Most home brewers tend to stockpile byproducts because they are uncertain of appropriate disposal methods. These large stockpiles of byproduct can potentially lead to a significant fire hazard. As risks exist, care should be taken by first responders when signs of a biodiesel facility are noted. Though many operations can legally produce biodiesel, they may still cause harm to emergency personnel. The MIOC bulletin offered the following considerations for first responders:

- Chemicals involved in production are legal for residential storage, but only in limited quantities.

- Methanol burns with an invisible flame.

- Methanol vapor can be released, causing poor air quality in a confined space.

- Most models of photoionization detectors will not detect methanol.

- Methoxide is a highly caustic chemical that has been associated with nerve damage caused by corrosive burns.

- Chemicals involved in production are flammable and can pose a significant fire hazard.
-

Overheated oils can add to a fire load.

- Use of pressure tanks in production can result in explosions.

3.15 Biodiesel production permitting considerations

> Every attempt has been made to provide the best possible guidance associated with biodiesel production, but due to the complex regulatory structure, local variable, and a rapidly evolving environment, definitive information cannot be given for every location

> Permission is extremely complex and can easily take 6 months or more

> Even if your facility is exempt from permitting requirements, it may be subject to complaint-based or discretionary enforcement

3.15.1 Oil collection

- Some cities regulate collection containers
- Used fryer oil is considered solid waste and will be treated as municipal waste for transportation purposes
- Some areas require chain of ownership documentation
- Spill protection gear
- Professional spill remediation
- Insurance

3.15.2 Zoning

- Biodiesel production falls under SIC code 2869 and NAICS 325199
- Consult with planning and zoning department ahead of operations, a conditional use permit may be an option
- Methanol will be an issue in town, a fireproof cabinet will be required and no more than 120 gallons in any single workspace. Workspaces must be separated by fire proof walls

3.15.3 Air quality and construction permits

- Air quality permit may be required for plants emitting more than 5 tons VOCs or NOX if in a non-attainment area

- Small non-commercial plants are usually exempt if they are not operating an internal combustion engine or boiler to power facility and are not part of a larger operation

- If no ground is being disturbed a construction permit usually is not required, but check with DNR, DEQ first

3.15.4 Operating permit?

- Some states may require an operating permit that is often based on the construction permit

- Some counties require an operating permit even if the state does not require a construction or air quality permit

- Some counties have no deminimus threshold, thereby requiring a permit for ANY VOC release

- On-farm production considerably reduces oversight

3.15.5 Wash water

- Wash water may not be put down the drain without written permission; most municipal waste water treatment plants cannot handle it

- Land application may be appropriate, but a permit must be obtained from DNR or DEQ or whatever agency has jurisdiction in your state
- Wash water chemistry (depends on

Methods

- BOD — 780mg/L
- Total Kjeidalh N — ranging from 9-15
- Nitrate + Nitrite — ranging from .1-6
- Total phosphorus — around 5.7
- Total potassium — around 970 mg/L

3.16 Glycerol

- It is very important to find a legitimate use for your glycerol so that it not is a hazardous waste but remains only a hazardous material.
- Storage for more than a year may trigger a hazardous waste determination
- Can be burned in a waste oil burner, but usually must be diluted; has a BTU rating around 95,000
- Has value as boiler fuel, but arrange for disposal
- Glycerol may be land applied, used for dust control or fed to livestock under certain conditions and has a similar feed value to corn

3.17 EPA and state regulation of storage tanks

- EPA requires Spill Containment, Control and Countermeasures (SPCC) plan for facilities with over 1320 gallons of combined storage

- The EPA regulates spills in waterways of 40 gallons or more and must be reported to the National Response Center along with a spill of Methanol over 758 gallons or 1,000 pounds of KOH or MeOH

- All storage tanks over 250-300 gallons (depending on the state) must have secondary containment

- These regulations change frequently; check with your state environmental agency

3.18 Dispensing issues

- Introduction of fuel into commerce
- ASTM 6751-9
- UL listing
- Tax Liability, Federal Letter of Activity M(637) and Form 720 to pay road use tax, due quarterly
- State road use taxes due monthly
- Renewable Identification Number (RIN) required for < 20,000 gallons per year production

3.19 Workplace safety
- **Ventilation**

- Your shop MUST be well ventilated
- The OSHA PEL (Permissible
- Exposure Limit) for methanol vapors is 200 ppm
- Ventilation fans should have sealed motors
- Methanol recovery can reduce risks

- **Wiring**

- Bad wiring, electrical may be your biggest danger!
- Prevent sparks, static
- Breaker your equipment
- Bonding and grounding is essential
- Hard wire equipment, use
- Switches, not plugs!

- **Fire**

Fuel	Flashpoint temperature
Gasoline	-40° F
Diesel	143° F
Methanol	54° F
Biodiesel	300° F

- Biodiesel IS NOT a fire hazard
- Eliminate static and sparks
- Install fire suppression/extinguishers
- Oil soaked rags or sawdust can spontaneously combust
- Communicate with the first responder

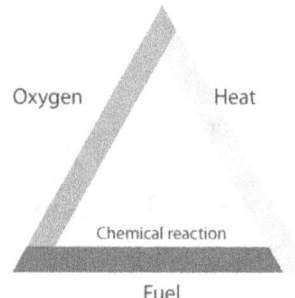

- **Spill containment**
  - Concrete floor with curb under equipment and storage
  - Degreaser and oil-dry
  - Check with state, what is "reportable amount"?
- **Compatibility with equipment**
  - Warranties
  - Fuel-related problems
  - Cold weather issues

- **Environmental safety**
  - All safety issues are environmental issues
  - Air issues
  - Water issues

3.20 Discharge to sanitary sewer

Discharge of pollutants to the sanitary sewer, including biodiesel washing, has the following restriction:

- The only waste product, from the production of biodiesel, allowed to be discharged to the sewer is the water used in the washing of the biodiesel.

Using the dry wash method is preferred.

- No permanent plumbing connection to the sewer is permitted from the biodiesel process.

- No discharge of any waste byproduct such as glycerin or bad product is allowed to be discharged to the sewer.

- For information on disposal of glycerin or bad product, call household hazardous waste or Greenfield wastewater treatment plant.

- Be sure to follow all recommended procedures and handling precision of the chemicals and equipment used.

3.21 Blending and storage of biodiesel

The biodiesel produced can be stored in its pure form or mixed with petroleum diesel. The mixing can be done by splash blending, injection blending or simply pouring the two products together and agitating. The standard storage procedures used for petroleum diesel can be used for biodiesel blends up to B20 (B20 is defined as 20% biodiesel and 80% petroleum diesel by volume). B100 or blended biodiesel higher than B20 should not be stored for more than 6 months, and it is recommended that all fuel be used within 90 days. The possible sources of contamination during storage include water, dust, microbes, spores, bacteria, algae, in addition to simple oxidation of the fuel. Therefore, the acid value should be monitored, and fuel-enhancing additives may need to be added.

3.22 Storage Tanks and Dispensing Equipment

Most tanks designed to store diesel fuel will store B100 with no problem. Acceptable storage tank materials include aluminium, steel, fluorinated polyethylene, fluorinated polypropylene, Teflon®, and most fiberglasses. If you are in doubt, contact the tank vendor or check the National Biodiesel Board (NBB) Web site Brass, bronze, copper, lead, tin, and zinc may accelerate the oxidation of diesel and biodiesel fuels and create fuel insoluble (sediments) or gels and salts when reacted with some fuel components. Lead solders and zinc linings should be avoided, as should copper pipes, brass regulators, and copper fittings. The fuel or fittings tend to change colour, and insoluble may plug fuel filters. Affected equipment should be replaced with stainless steel, carbon Steel, or aluminium.

In some locales, an Underwriters Laboratories (UL) listing is required by insurance companies or by state or local Regulations for the tanks used for fuel storage and for equipment used to dispense fuel. (UL is an independent, Not-for-profit, nongovernmental organization that tests products for public safety.) As biofuels use has become more widespread, the issue of UL listing has become a larger concern as UL-certified equipment for biodiesel and biodiesel blend service is extremely limited. Vendors must submit examples of their equipment to UL for testing.

UL testing programs are currently underway for equipment to be used with biodiesel and biodiesel blends, including dispensers, aboveground steel tanks, underground steel tanks, nonmetallic tanks, aboveground and belowground piping, coatings, sumps, and heating equipment.

However, until UL certification is available, the lack of permissible equipment may be a barrier to biodiesel and biodiesel blend storage and use in some locations.

3.23 Materials Compatibility

B100 will degrade, soften, or seep through some hoses, gaskets, seals, elastomers, glues, and plastics with prolonged exposure. Some testing has been done with materials common to diesel systems, but more data are needed on the wide variety of grades and variations of compounds that can be found in these systems, particularly with B100 in

U.S. applications. Nitrile rubber compounds, polypropylene, polyvinyl, and Tygon® materials are particularly vulnerable to B100. Before handling or using B100, ask the equipment vendor or OEM if the equipment is suitable for B100 or biodiesel. In some cases, the vendor may need the chemical family name for biodiesel (the methyl esters of fats and oils) to look up the information or even the exact chemical name of some of the biodiesel components, such as methyl oleate, methyl linoleate, methyl palmitate, or methyl stearate. Oxidized biodiesel and biodiesel blends can contain organic acids and other compounds that can significantly accelerate elastomer degradation. (Published data on B100 material compatibility are summarized in Appendix E.) There have been no significant material compatibility issues with B20, unless the B20 has been oxidized.

If your equipment is not compatible with B100, the materials should be replaced with materials such as Teflon, Viton, fluorinated plastics, and nylon. Consult B100 suppliers and equipment vendors to determine materials compatibility, and ask B100 vendors in other regions what problems they have experienced and what kind of replacement materials they are using. It is advisable to set up a monitoring program to visually inspect the equipment once a month for leaks, seeps, and seal decomposition. It would be wise to continue these inspections even after one year, as experience with B100 is still relatively limited.

3.24 Cleaning Effect

Methyl esters have been used as low-VOC (volatile organic compound) cleaners and solvents for decades. Methyl esters make excellent parts cleaners, and several companies offer methyl esters as a low-VOC, nontoxic replacement for the volatile solvents used in parts washers. Because B100 comprises methyl esters that meet ASTM D6751, it will dissolve the accumulated sediments in diesel storage and engine fuel tanks. These dissolved sediments can plug filters. If this happens, it can cause injector deposits and even fuel injector failure. If you plan to use or store B100 for the first time, clean the tanks and any parts in the fuel system where sediments or deposits may occur before filling the tanks with B100.

The level of cleaning depends on the amount of sediment in the system (if the system is sediment-free there should be no effect) as well as the biodiesel blend level (the higher the blend level, the greater the cleaning potential).

The cleaning effect is much greater with B100 and blends with 35% or more biodiesel, in comparison to B20 and lower blends.

Biodiesel spills should be cleaned up immediately, because biodiesel can damage some types of body and engine. Biodiesel can also remove decals from tanks or vehicles near fueling areas. All materials that are used to absorb biodiesel spills should be considered combustible and stored in a safety can.

3.25 Microbial Contamination

Biocides are recommended for conventional and biodiesel fuels wherever biological growth in the fuel has been a problem. If biological contamination occurs, water contamination should be suspected and will need to be controlled because the aerobic fungus, bacteria, and yeast HC-utilizing microorganisms usually grow at the fuel water interface. Anaerobic colonies, which usually reduce sulphur, can be active in sediments on tank surfaces and cause corrosion. Because the biocides work in the water phase, products that are used with diesel fuels work equally well with biodiesel.

3.26 Codes and Regulations

Biodiesel blends are subject to the same regulations and codes as diesel fuels. Blends up to B5 are considered regular diesel and approved for use in existing diesel infrastructure. Blends above B5 are subject to additional requirements. This section focuses on the federal requirements. Stations considering blends above B5 should contact their state and local authorities to identify other regulations and requirements. EPA's Office of Underground Storage Tanks regulates underground storage tanks (USTs) per Code of Federal Regulations, Title 40, Subtitle 1 Subchapter 1 Parts 280-282.The federal UST regulation was updated on October 2015 with section 280.32 in the 2015 UST regulation providing clarity to the 1988 compatibility requirement by specifying additional compatibility requirements for owners and operators wishing to store certain regulated substances, including diesel containing more than 20%. All portions of a UST system must be compatible with the fuel stored. Demonstrations of compatibility must be provided for the tank, piping, containment sumps, pumping equipment, release detection equipment, spill equipment, and overfill equipment.

The requirements are:

1. Owners of USTs switching to store blends containing greater than B20 must notify their implementing agency (usually a state office) 30 days prior to switching fuels to store B20+ blend.

2. Owners of USTs storing blends greater than B20 must demonstrate compatibility through either:

    a. Certification/listing of equipment for use in the fuel stored by a nationally recognized, independent testing laboratory, or

    b. Equipment or component manufacturer approval for use in the fuel stored. This statement must be in writing affirming compatibility and must list the specific ranges of biofuel blend the equipment or component is compatible with, or

c. Use of another option determined by the implementing agency to be no less protective of human health and the environment.

3. Owners of USTs storing blends greater than B20 must maintain records demonstrating compatibility as long as that fuel is stored.

OSHA regulates some fuel-dispensing equipment. Its regulations applicable to service stations have not been updated in decades and therefore do not specifically address biofuels. OSHA is planning to update these standards to address new fuels in the marketplace. OSHA 1910.106 (g) (3) (IV) and (g)(3) (VI) (a) require dispensers and nozzles to be listed by a third party for specific fuels.

OSHA 1910.106(b)(1)(i)(b) and (c)(2)(ii) requires tanks, piping, valves, and fittings other than steel to use sound engineering design for materials used; however, there is no listing requirement. OSHA 1910.106(b)(1)(iii) covers steel tanks and requires sound engineering and compliance with UL 5825 and American Petroleum Institute Standards 65026 and 12B27 as applicable.

Local authorities have jurisdiction typically adopt fire codes from one of two organizations, the National Fire Protection Association (in particular, Code 30A, 28 which includes language on alternative compliance to address new fuels) or the International Code Council, which provides standard codes for retail stations that are accepted or modified to meet local requirements. Other organizations developing best practices and codes include The American Petroleum Institute, Fibreglass Tank and Pipe Institute, NACE International, National Conference on Weights and Measure, National Leak Prevention Association, Petroleum Equipment Institute, and Steel Tank Institute.

4 Setting up the workshop

Design and construction of a small-scale biodiesel plant.

4.1 Process Design

The specific goals for the design are:

- Minimize manual labour

With only one employee, the more automation in the system the better. Pumping systems were designed to eliminate the need for manual handling of products from one stage to another.

- Minimize costs

With a tight budget, cost has been always the top concern. While there was a need to purchase essential tools, simplicity and cost-prevention measures remained a constant concern.

- Flexibility

The piping system was designed for one possible processing method, but several other methods were also identified. It is important that NativeSun conserves the ability to experiment with different methods.

- Equipment limitations

Two pumps were purchased for the process. Meeting all pumping needs using only 1 or both of the pumps thus became a significant hurdle to overcome. Piping and valve systems were designed to ensure the two pumps could perform all required tasks without needing to be moved or reconnected.

The process for making biodiesel is summarized as follows:

- Preheat used vegetable oil to remove water
- Determine oil pH and measure appropriate quantity of NaOH and methanol
- Mix reactants
- Allow glycerin to settle out, drain glycerin and send to methanol recapture
- Wash biodiesel to remove trace contaminants
- Set in the sun to drive off traces of water

The process design is divided into 4 stages; the pre-heat, the processing and the wash stages that were previously part of the NativeSun process, as well as a methanol recapture stage which is a new addition. Also explored were a water treatment stage and solar heating system.

4.2 Preheat Stage

The role of the preheat stage is to remove any water that might be mixed with the used vegetable oil. Water is detrimental to the transesterification process and must be removed as much as possible. Heating the oil to 50-54 OC is enough to cause the water to settle out.

There are two preheat tanks, both are used tanks purchased from the Bico ice cream factory in St. Michael. The tank volumes are 150 gal. Using two tanks allows for the continual receiving of used vegetable oil. Hot water flows from a propane heater and through a heating coil to heat the oil, the cooled water is then returned to the heater.

The valve system is designed such that hot water will flow through only one tank at a time, likewise the vegetable oil will be pumped out of the tanks sequentially. While one tank is pre-heating, the other is available for receiving used oil.

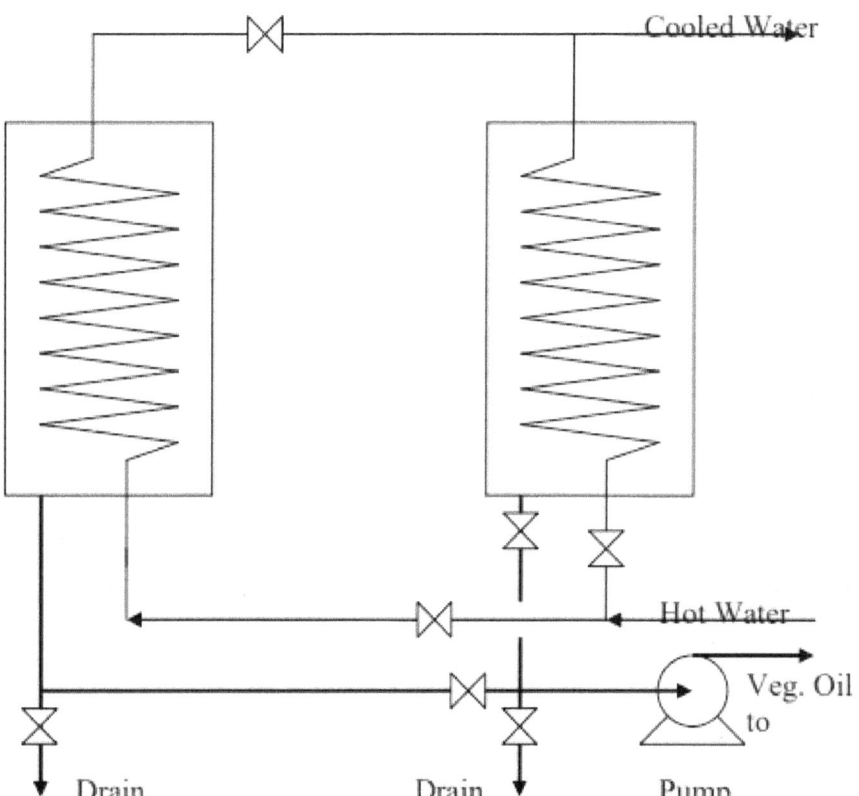

4.3 Processing Stage

The processing stage consists of two reactor tanks, each connected to its own pump for recirculating and mixing of the reactants. The first tank, reactor A, is tall and narrow with a conical bottom. This reactor shape provides adequate mixing of the fluid when recirculated by the pump. The second tank, reactor B, is bulkier and rounded on the bottom, which would create dead zones if mixed by simple recirculation. To overcome this problem and ensure adequate mixing in both processing tanks, a mechanical mixer with impellers was purchased. Static mixers were also purchased to increase mixing within the pipes.

The goal of the processing piping system is to ensure flexibility of the system while maximizing use of only two pumps. The primary feature of this system is that it allows for the transfer between the two reactors. Used vegetable oil can be pumped into either tank via pump A, and then recirculated with the reactor tanks via their respective pumps. The Contents can be transferred between the two tanks via this piping scheme. The proposed processing method is outlined as follows:

- Used vegetable oil is pumped from the preheat stage into reactor B.
- Mixed methoxide is added to reactor B and circulated to provide initial mixing.
- Half of the mixture is then transferred to reactor A, and both tanks are recirculated for the required reaction times.
- Following the reaction, the tanks are allowed to settle and glycerine is drained out for eventual methanol recapture.
- The biodiesel product is pumped out towards the washing stage.

Initially mixing all of the reactants in one container ensures homogeneity of the oil: methoxide ration within both reactors. This removes one measuring step. Using both reactors for processing allows for eventual expansion. According to the business plan, NativeSun projects to initially process 150 gal /week, increasing production towards a

300 gal/week. The system also maximizes the use of the two pumps, which perform all the required transfers and recirculating.

Pump A is significantly larger than pump B. It also has more power and the additional feature that it can run dry without damaging. These two features make pump A better choice for pumping used vegetable oil into the reactor, as it doesn't require the operator to monitor the level in the preheat tanks.

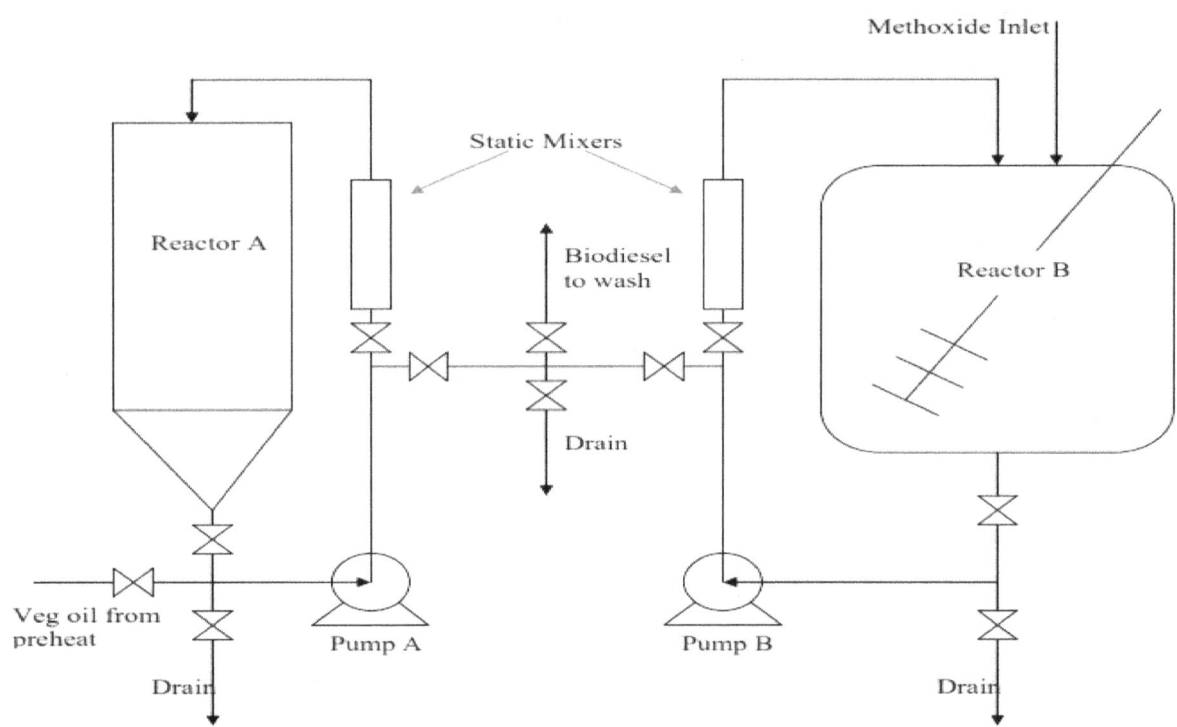

Schematic of Processing Stage

4.4 Washing Stage

There are many methods of washing available when processing biodiesel. An ideal washing method maximizes interaction between the fresh biodiesel and washing water, while minimizing the risk of emulsifying2. Previously, NativeSun had simply mixed water and biodiesel manually. This works well for small batches, however the manual aspect makes it increasingly time-consuming with increase in batch size, and more difficult to control to avoid emulsification if automated.

The most commonly recommended method of washing biodiesel is aeration, whereby air is bubbled through layers of water and biodiesel. The bubbles rising into the biodiesel carry with them a thin film of water. The biodiesel contaminants dissolve in the water. When the bubble bursts, this water falls out of the biodiesel layer and returns to the water layer, bringing the contaminants with it. This method yields a very high water-biodiesel surface area and minimizes the risk of emulsification.

The wash tank design makes use of the following items we had at our disposal:

- A 1 m3 plastic tank donated by a paint company.
- The compressor required for the operation of the pneumatic pumps.
- Lengths of ½" PVC piping for the aerator.

The aerator was constructed by drilling small holes along the PVC pipe at 3" intervals. Fourlengths of tubing, running parallel along the bottom of the tank, are connected to a single shaftextending to the top of the tank. The compressor connects to a regulator and then to the shaft. Therate of air flow through the aerator, and thus the rate of bubbles, is controlled by the regulator.

For each batch, an amount of water equal to half the volume of biodiesel is added to the wash tank,followed by the biodiesel. The product is bubble-washed for 6 hours and then let to settle for 1hour. This wash is repeated three times, each time the water is drained from the bottom and newwater is added.

One identified problem with the aerator system is that biodiesel degrades PVC glue over time. As such, it is important to minimize the interaction between the aerator and the biodiesel. The paint tank has a built-in drain located at the bottom, but an additional tap was installed a few inches higher, which is the tap that will be used to drain the water to ensure that a certain volume of water will remain above the aerator at all times, protecting the PVC piping.

The total volume of the wash tank is 350 gal. Given that some airspace is required at the top, and 1/3 of the tank's volume must be water, this wash tank cannot wash biodiesel batches greater than 200 gal. Should NativeSun choose to increase production, a second wash system would be required.

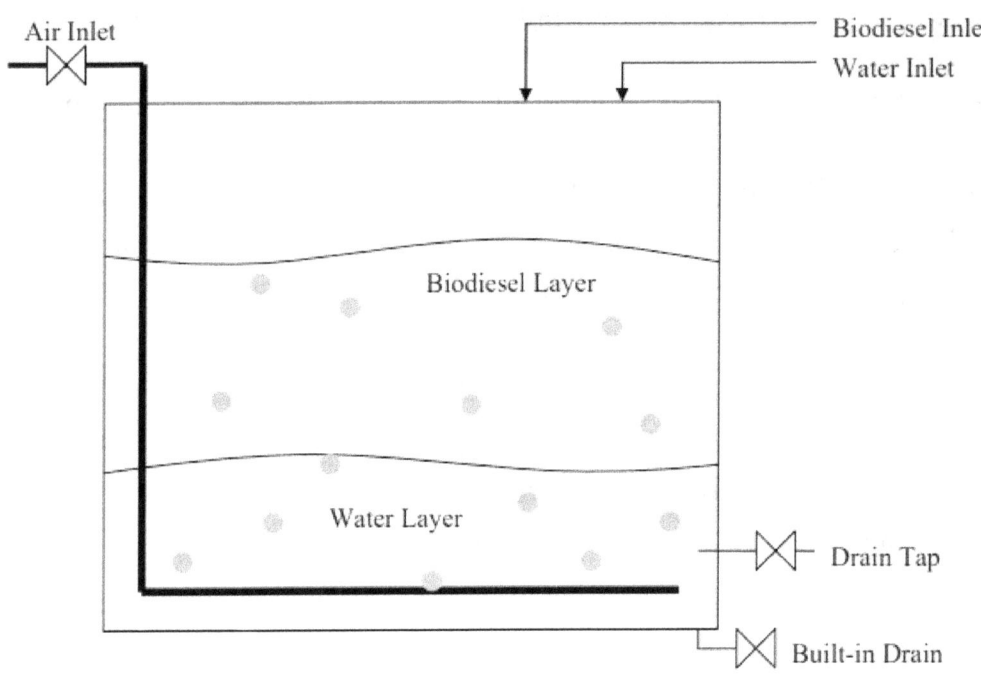

Schematic of Wash Tank

4.5 Methanol Recapture

Methanol is the most expensive input for the biodiesel process. As methanol is derived from petroleum, it must be imported into Barbados resulting in an inflated cost. The price is also tied to the price of oil, and as such is at risk of instability as crude prices rise.

In order to ensure that the reaction consumes all of the vegetable oil, methanol is added in excess to force the equilibrium to the right. This excess methanol ends up in the glycerin by-product and represents a significant loss. Methanol in the glycerin also limits the potential for marketing it as a product, as the combination is deemed unsafe and flammable.

Given the relatively low boiling point of methanol, it is possible to recapture the methanol via a simple still. The mix of glycerin and methanol, still liquid following the reaction stage, can be heated to vaporize the methanol. These vapors can then be condensed and recycled, maximizing use, reducing waste, and lowering overall processing cost.

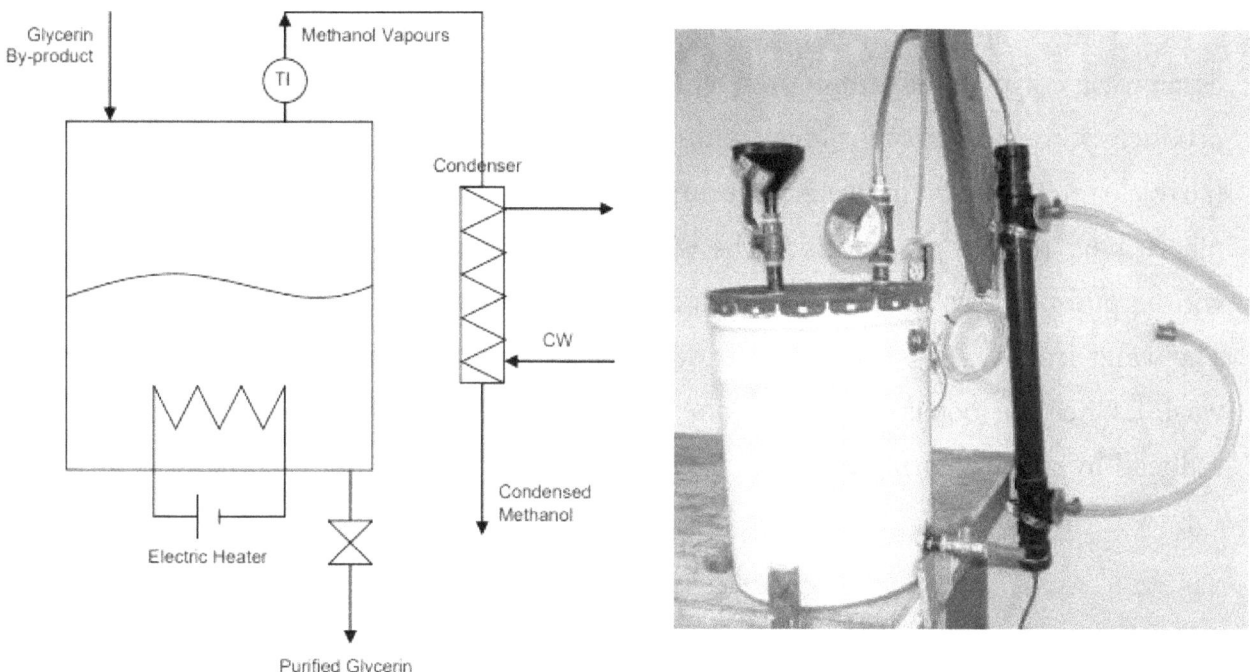

Schematic of Methanol Recapture Unit

The glycerin by-product is poured into a chemical container through a funnel in the lid. The container must be sturdy and airtight. An electric heater heats the glycerin to the methanol boiling point of 66°F. The vapors rise through the bucket and into a length of copper tubing. The copper tubing, then coils as it enters a condenser. Cold tap water passes through the condenser, cooling the methanol vapors of a liquid. The liquid falls through the copper tubing where it is collected at the bottom. Once the liquid methanol stops flowing, the process is complete and the glycerin is drained from a tap while still liquid.

4.6 Water Treatment

In an effort to reduce the overall impact of biodiesel production, water consumption and recycling were considered. Three wash stages are required for biodiesel production, the first stage, taking out the most contaminants and each subsequent wash containing significantly less. Water recycling then becomes an ideal method for reducing overall water consumption within the process.

Each wash stage, at maximum production capacity uses 100 gal of water. The first wash produces water too dirty for reuse, but the water from the second and third washes are fairly clean. To reuse the water, two interim storage tanks of 100 gallons each will be placed near the washing station. The water from the second and third washing stages will be pumped into the storage tanks and then reused in the following wash process. The water from wash 3 will be reused for wash 2, and the water from wash 2 reused in wash 1. Once the loop is established, the required water consumption for washing will be reduced by 2/3, requiring only half a gallon of water per gallon of biodiesel instead of 1.5.

4.7 Outline of Water Recycling Process

In addition to water recycling, the following water treatment options were also explored, with the hopes of bringing the process water demand to nearly zero.

4.7.1 Solar Distillation

This method of water purification produces almost perfect distilled water and would remove all of the glycerin, methanol, and dissolved solids. Solar distillation, however, is highly inefficient and would require an extremely large surface area in order to treat the amount of water needed by NativeSun for biodiesel production. Through some research and testing, it was determined that a solar still covering an area of 4m2 could only treat about 30 gallons a week, simply not enough to meet the needs of NativeSun, and anything larger would take up too much space and be too costly to implement.

4.7.2 Slow Sand Filtration

Slow sand filters are very efficient at removing organic solids and quite inexpensive to construct. The problems with this type of system are that it may clog up if the waste water is too turbid, and it will not remove dissolved solids. This means that the waste water from the washing stage of biodiesel production might have to be pre-treated to decrease turbidity in order to pass it through a sand filter. Furthermore, it remains unclear exactly how much dissolved caustic soda would remain in the treated water and its effects on whether the water would be acceptable for re-use. Finally, there may be a problem with using a slow sand filter for most of the actual water treatment is performed by the microorganisms living in the top layer of the sand and it is still unclear how these organisms would fair with the filter not in constant use. Unfortunately, there were no funds available to construct a test sand filter.

4.7.3 Methanol Evaporation

In some of the literature examined for the biodiesel internship there was mention of heating the unwashed biodiesel to boil off the methanol and cause the suspended solids to precipitate out. The theory is that methanol evaporates at a much lower temperature than biodiesel (only 60C) and thus could safely be evaporated out which would release the suspended solids leaving clean, useable biodiesel. This step would, in theory, altogether eliminate the need for a wash stage and thus eliminate the need for clean water. The internship group was able to perform some basic tests on this theory which were however inconclusive. Due to time constraints there was no further experimentation, but this process definitely merits more thought and testing.

4.7.4 Solar Heating

In further attempts to reduce the environmental impact of biodiesel production, solar heating was investigated. The energy requirements for the oil pretreatment stage could easily be met using a solar heating system. We have met with Mr. Vincent MacLean, CEO of Aqua Sol, who generously offered technical support in designing such a system.

The system is a very simple one where a solar unit installed on the roof at the Future Centre would be connected to the heating coil of the preheating tank. The design requires a pump to allow the working fluid to circulate through the element up towards the solar panel. A debate emerged between two possible options: indirect or direct solar heating. Direct heating involves having the used vegetable heat directly as it flows through the coil of the solar panel. However, this option was eliminated as it causes tube-side fouling hence resulting in higher maintenance needs. We opted for indirect heating where water flows through the system.

4.8 Obstacles and Problem Solving

This section will detail some of the major obstacles that were encountered throughout the process and the attempts to resolve them. The obstacles are loosely divided into technical, and equipment. While the list is not exhaustive, it does serve to highlight some of the major issues dealt with.

4.8.1 Technical obstacles

- **Installing the Electric Mixer**

Adequate mixing of the vegetable oil with the methoxide is very important in the biodiesel process. As mentioned above, Reactor B, with its ellipsoid shape does not provide the vortex required for adequate mixing. To alleviate this problem, an industrial strength electric mixer was purchased from the United States. The plan was to have this mixer mounted on Reactor B. When the mixer finally arrived, several weeks late, it became apparent that this would be no small feat. The mixer itself weighs approximately forty pounds, and is very awkward to handle.

The proposed solution is to weld two brackets to the tank and secure the mixer to them with a steel bolt. These brackets would need to be steel, at least ¾ inches thick, and would also need to follow the curvature of the top of the Reactor. To accomplish this, a cardboard outline of the curvature was created and the brackets were cut by Williams Steel on a CNC machine at no charge using steel.

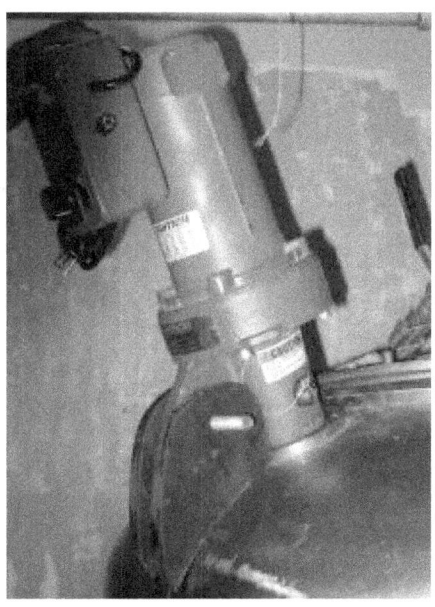

The electric mixer with the mounting brackets

The brackets have yet to be welded to Reactor B, as we require a welder to come to perform this welding.

- **Leaking Pipes**

One of the major concerns with the main reaction stage was pipe leakage. Using a compressor on loan from Williams Industries, we tested for leaks by pumping water through the network. Leaking was observed in several nodes of the network. Since oil is more viscous than water, the amount of leaking in the piping will be far less when the oil is being processed. Nonetheless, it is desirable to eliminate all possible leaks.

Many of the leaks were removed by tightening the connections and sealing them with plumbers tape. However, the unions (a pipe connection that allows pipes to be connected in situations where tightening one connection loosens another one) proved very resistant to sealing. These connections are prone to leaking because the interface between the two halves of a union is a metal to metal connection. It is impossible to add plumbers tape to this interface because there is no surface to wrap the tape around. Instead, unions are usually sealed by tightening them to such an extent that the steel surface on one faces bites into the softer surface on the opposite face. However, our purpose for the unions was to allow easy disassembly of the pipes for cleaning and maintenance, and to repeatedly tighten and loosen the unions would quickly damage the metal-on-metal seal.

A leaky union in the network

An alternate method of sealing unions is through the use of rubber O-rings. These O-rings act as gaskets and, when squeezed between the two metal surfaces of the union, provide an effective seal. O-rings, available at the hardware store, were installed in the 1" unions. However, the 2" unions, located at the entrance and exit of Reactor B, are too large for standard O-rings. To remedy this, gaskets were constructed out of cork sheet. Both gaskets were effective at removing leaks in the network.

The only remaining source of leaking is at the exit of Reactor B. Reactor B does not have a standard pipe threading on it, but it is possible to connect it with a 2" thread. Plumbers tape has been unable to seal the connection. Silicon caulking was applied to attempt to eliminate the links, but the effect has yet to be verified.

- **Measuring Flow and Fluid Levels**

One of the difficult aspects of the main reactor stage is that it is essentially a black box. Once the reactants are in the reactors, there is no viable way of measuring what is happening. Knowing the flow rate through the mixers is very important for process control. Furthermore, a flow-meter would also be helpful in determining how much fluid is left in the tanks when pumping out of the reactors and into the wash stage. Transparent flexi-piping has been installed which gives a visual approximation of fluid flow, but not actual quantifiable measurement. The four-way valve with theplastic tubing is shown in Figure 10. A flow-meter would be a useful tool should funds become available

The 4-way valve with the height-gage attachment

With regards to monitoring fluid levels in tanks, sight gages were installed for both tanks. Clear PVC tube is connected to the bottom of the each tank via a valve. By opening the valve, the level inside the tank is indicated within the tubing. This will help ensure that the pumps will not run dry when draining the tank, as well as help when pumping roughly half the biodiesel reaction mixture from reactor B to reactor A. It was felt that this was a cost-effective solution to the problem.

4.8.2 Equipment Obstacles

- **Pumps**

Due to economic constraints, the two main reactors differ in both shape and size. These differences impact the mixing abilities of the tanks. Whereas the conical bottom of Reactor A causes a vortex that mixes the biodiesel, Reactor B has an ellipsoid shape and does not form the vortex. To compensate for the reduced mixing, Reactor B will have an electric mixer as well.

The two Reactors

In the current piping network, each reactor is coupled with a pump. Both pumps that were purchased run on compressed air. This is a safety consideration as electrical pumps sometimes spark, which could ignite the biodiesel. Furthermore, NativeSun NRG chose a larger, more powerful pump for Reactor A. The reason for the larger pump is that Reactor A is higher, and since the critical mixing in Reactor A is due to the vortex, it is imperative that the pump be powerful enough to create one.

The first obstacle encountered concerned the pump connected to Reactor B. The pump did not arrive with an exit flange. Pumps rarely come with a flange because each flange can only accommodate one size of pipes, and there are too many pipes sizes for the manufacturer to supply flanges for. Instead, flanges are usually purchased at specialty stores. Without a flange the pump could not be connected into the network. We were unable to find any stores on the island that had the required flange. Fortunately, we were able to have flanges machined for us by technicians working for the Barbados Water Authority.

A second obstacle was encountered upon examining the documentation supplied with the pumps. Both pumps are more powerful than required. This is problematic as pumps are designed to operate at a certain capacity. If they run too far below capacity, there is a risk they will stall. To ensure that they would not stall, the pumps were tested by with water in the network. Both pumps operated satisfactorily; however, there is still concern that operating the pumps outside of their design range will have long-term consequences.

- **Compressor**

As has been previously mentioned, a compressor is required to operate the pumps and the bubble washer. Before purchasing a compressor, however, it needed to be sized. The sizing process was difficult due to two different, and over-designed, pumps running concurrently on one compressor.

To help size the compressor we contacted the pump manufacturers, as well as Williams Industries, and two consultants from the BIDC. Eventually a gasoline-powered compressor was chosen. This compressor was tested for a day by running water through the network. There was not enough water in the tanks to run both pumps simultaneously, so only one pump was tested at a time. The compressor was able to fulfill the pressure requirements of each pump. It is believed that the compressor storage tank will be capable of supplying enough airflow for both pumps to run simultaneously, and NativeSun NRG intends on confirming this before purchasing one.

The lack of a compressor caused a serious delay in the project. Without one, no tests on the piping could be performed and no processing could be attempted. Much time and energy was invested in finding a compressor to borrow for a day in order to test the equipment.

- **Static Mixers**

To help with the mixing process in the main reaction, static mixers were purchased. However, the mixers are made from PVC and they may not be able to withstand the strain of pumping biodiesel. For example, the threading on Mixer B was cracked when

Pump B accidentally fell over. While new threading was installed, this highlights the safety concern for the mixers.

This shows the mixer attached to Reactor A. As can be seen, both ends of the mixer are connected to steel piping components. Since steel is far stronger than PVC, any stresses placed on the network will deform the mixers first.

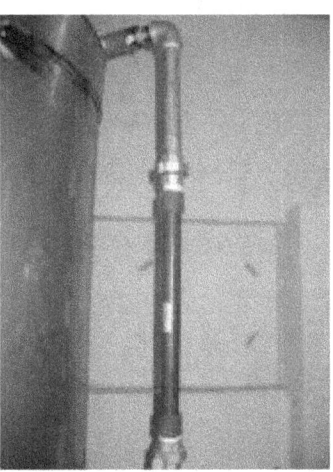

The static mixer connected to Reactor A

Two different methods were proposed to secure the mixers. The first solution involved creating a cage around each static mixer by welding bars onto the unions that connect the mixers at both ends. There was concern that the heat from the welding would melt the mixer. Furthermore, once welded, it would be impossible to remove the mixer from the cage. A second proposal involved running a parallel pipe next to each static mixer. This pipe would be closed off to fluid, the only purpose being to bear the stresses place on the structure. It would not require much extra plumbing equipment to implement such a parallel system. Due to financial constraints the solution was never implemented, however this can be performed quite easily in the future.

- **Pipe Threading**

One obstacle that arose during our work was the threading of the pipes. In the past, NativeSun NRG used both PVC and flexible pipes in the operation. Neither PVC nor flexi-pipe requires threading as the PVC is glued and flexi-pipe is bracketed. However, for long-term durability in the plant, steel pipes were chosen as the primary mode of piping.

Steel pipes are purchased in a standard length of twenty feet. Before they are used they must be cut and threaded. Cutting the pipes poses no problem, as they can be cut with a hacksaw. Threading the pipes requires special equipment that we lacked, there was a significant time factor involved, as it would sometimes take several days before the pipes could be threaded.

5 Chemistry of biodiesel

5.0 Base Transesterification

Base-transesterification involves using an alcohol (i.e. methanol, ethanol) and a catalyst (either sodium hydroxide or potassium hydroxide) for the conversion of a triglyceride (i.e. waste vegetable oil or animal fat) into an alcohol-ester (biodiesel). Glycerin is a byproduct of the reaction.

In addition, soap may be formed from the reaction between catalyst (sodium or potassium hydroxide) and free fatty acids present or released in the waste vegetable oil. Alcohol-esters are also known as "biodiesel", and are named according to the alcohol used in the reaction. For instance, if methanol is used for processing, the resulting ester is classified as a methyl-ester; if ethanol is used for processing, the resulting ester is classified as an ethyl-ester.

A simple conceptual model of an ideal base-transesterification reaction follows:

$$\underbrace{\begin{matrix}\text{ESTER} \\ \text{ESTER} \\ \text{ESTER}\end{matrix}\!\!\!-\text{GLYCERIN}}_{\text{TRIGLYCERIDE}} + \underbrace{\text{METHANOL}}_{\text{ALCOHOL}} \underbrace{\Longrightarrow}_{\text{NaOH} \atop \text{CATALYST}} \underbrace{\text{METHYL-ESTERS}}_{\text{BIODIESEL}} + \underbrace{\text{GLYCERIN}}_{\text{BYPRODUCT}}$$

In its simplest form, home brewing requires that you first dissolve a predetermined amount of catalyst in methanol to produce a solution known as methoxide. You then mix the catalyst methanol solution (methoxide) with your waste vegetable oil for the purpose of transesterification, or conversion of the vegetable oil to biodiesel and glycerin.

Left: Dark-colored glycerin byproduct drain from a biodiesel processor

Right: Light-colored biodiesel follows glycerin drainage from a biodiesel processor

Transesterification is not limited to vegetable oil; animal fats can also be converted to biodiesel. Please note that the cold flow properties of fuel rendered from animal fat will be poorer than that of fuel rendered from waste vegetable oil. As such, you probably want to avoid using oils that are contaminated with significant quantities of animal fats. For demonstrations purposes only, below are pictures of a simple experiment

100 ml of animal fat collected from bacon

summarizing how biodiesel can be rendered from bacon grease.

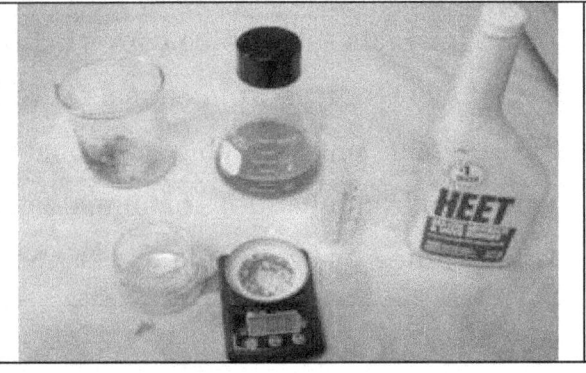 

The photo on the left shows warm bacon grease transferred to a flask; also note the ingredients for preparation of a small quantity of sodium methoxide. The photo on the right shows the addition of the methoxide (20 ml of methanol containing 0.60 gr of dissolved sodium hydroxide) to the flask containing warm bacon grease.

After mixing manually for at least one minute, the mixture is allowed to settle. Over time, the solution will separate into a layer of unwashed biodiesel (methyl esters) separated from a darker phase made up of mostly glycerin.

Note that unwashed biodiesel in this experiment may be contaminated with unreacted triglycerides, residual methanol, and soap. The darker glycerin phase will also be contaminated with residual catalyst, residual methanol, and free fatty acids. More details regarding these contaminants are presented in the following section..

The details of base transesterification described up to this point have been greatly simplified. Realistically, the reactions associated with home brewing are more complicated. Oil impurities, processing time, and non-ideal reactant masses may also result in the formation of additional byproducts. The following summarizes some of the considerations which may impact the biodiesel rendering process.

> ➢ Acid content of oils: Used and/or degraded oils will often contain high levels of free fatty acids (FFA). If acidic oils are used for the production of biodiesel, FFAs in the oil will consume catalyst and prompt the formation of soaps. If the base catalyst is consumed in making soap, additional catalyst must be introduced to ensure it is available for the transesterification of triglycerides to biodiesel. As

such, the acidity of the oil used as a feedstock will have a direct impact on the amount of catalyst needed, and will also contribute to unwanted soap formation.

- Water content of oils: While transesterification is taking place, the presence of any water in the system causes a side reaction which begins with the hydrolysis of glycerides. Specifically, water in the reaction "breaks" a fatty acid chain from the glyceride molecule. As soon as the fatty acid chain breaks, it becomes a FFA. This raises the acidity of the oil which in turn consumes catalyst and promotes the formation of soap. As soap is formed, the water molecule is released back into the solution where it immediately attacks another glyceride.

This side reaction continues to make soap until all the catalyst in the solution is completely consumed. Fortunately, the hydrolysis reaction is relatively slow compared to the transesterification reaction, so only a small percentage of catalyst is consumed.

However, if the catalyst is consumed to make soap, it is now no longer available for transesterification. As such, the presence of water not only contributes to soap formation, it may also result in unreacted or partially reacted oil. For these reasons, a home-brewer should always use dry oil. Elimination of water in the oil is one of the few operations a home-brewer can do that has such a large positive effect on his finished product.

- Reactant masses: In order to ensure that the reaction moves forward to completion, excess methanol is typically used for the reaction. After transesterification, some of that excess methanol will partition to the biodiesel phase, but most will end up in the glycerin byproduct. If insufficient quantities of methanol and/or sodium hydroxide are used, or if sodium hydroxide is consumed through the creation of soaps, the base-transesterification reaction may not go to completion. This will result in a smaller yields of biodiesel coupled with the formation of mono and diglycerides, as well as unreacted triglycerides.

The presence of unreacted or partially-reacted triglycerides may be addressed through reprocessing.

❖ Mixing: Note that even if you begin with dry oil, there will always be water in the methanol-catalyst (methoxide) mixture since water is chemically formed when the catalyst is dissolved in the methanol. Proper mixing of your waste vegetable oil with the methoxide helps reactants overcome kinetic barriers for transesterification. Since the transesterification reaction takes place at a faster rate than the hydrolysis of glycerides, mixing helps promote the catalyst role for biodiesel production. Starting a reaction and letting it run overnight, or starting a reaction, then stopping it, then restarting several hour later are not good practices, and may result in a smaller yields of biodiesel coupled with the formation of mono and diglycerides, unreacted triglycerides, and greater amounts of soap.

5.1 Acid Esterification

Typical misting setup for washing biodiesel

Soapy wash water collected from mist-wash

Given the tendency for highly acidic oils to produce soaps, compromised oils may be difficult to process using a single stage base- transesterification. Large quantities of soap byproduct require multiple washings, and also requires disposal of the soapy wash water. This is time and resource intensive.

Of greatest concern, soaps can lead to the emulsifications during washing. When emulsifications do occur, they can be difficult to break, and require additional investment of time and materials. As such, the removal of soaps

98

requires the delicate introduction of water through misting. This is also known as mist-washing.

Acid esterification is employed on oils that are too acidic for simple base-transesterification chemistry. Acid esterification should be employed on oils that have an acid number greater than 6 as determined by a base titration. Acid number is determined through a simple titration with a 0.1% solution of sodium or potassium hydroxide, the details of which are discussed in the source-oil characterization phase of this document.

Acid esterification involves using an alcohol (i.e. methanol, ethanol) and an acid catalyst (i.e. sulfuric acid) for the conversion of free fatty acids present in acidic oils into alcohol-esters. A conceptual model of an ideal acid-transesterification reaction follows:

$$\underbrace{FFA}_{\text{FREE FATTY ACIDS}} + \underbrace{METHANOL}_{\text{ALCOHOL}} \xrightarrow{\underbrace{H_2SO_4}_{\text{CATALYST}}} \underbrace{METHYL\text{-}ESTERS}_{\text{BIODIESEL}}$$

Once acid esterification has converted FFAs to alcohol-esters (biodiesel), the resulting solution can undergo a base-transesterification. Base-transesterication will not affect the quality of any biodiesel generated during acid-transesterification reaction. Thus, the products from an acid esterification reaction may be directly incorporated as reactants in a base-transesterification reaction.

The benefits of acid esterification are twofold. First, free-fatty acids are converted to biodiesel. This causes overall biodiesel-yields to increase. Second, the conversion of free-fatty acids to biodiesel helps minimize soap formation during the base-transesterification process. This reduces the need for washing once the base-transesterification reaction is completed.

Realistically, the reaction is more complicated. Water may be formed as a byproduct of the acid esterification, and needs to be separated before executing a base-transesterification on the products of the reaction. In addition, not all free-fatty acids will be converted to biodiesel. As such, producers must recalculate the acidity of the acid-transesterification products and adjust the catalyst for base-base transesterification accordingly. Experienced small-scale biodiesel producers may see acidity drop from 12% to 5% when pretreating acidic oils through acid esterification.

5.2 Expected Yields

So, all this begs the question: "How much unwashed biodiesel and glycerin might I expect to yield from oil collected from my local restaurant?" Before I attempt to answer that question, let me clarify the following paragraphs in that when I say "glycerin", I'm really talking about the glycerin phase which will also contain residual catalyst and methanol as well as free fatty acids; and when I say "biodiesel", I'm really talking about unwashed biodiesel.

When employing base-transesterification of oils of an acceptable quality, I typically generate a volume of "glycerin" which is slightly less than the volume of the methoxide added to my processor. In other words, if I mix 35 gallons of waste vegetable with 7.7

Photo of sample collected from processor immediately after mixing.

The base-transesterification recipe used to generate this sample was 35 gallons of WVO (having a titration: 3.0 ml 0.1% NaOH for acid neutralization), 1060 grams NaOH and 7.7 gallons methanol. These constituents were mixed in an appleseed processor for three hours at a temperature of about 125 degrees Fahrenheit.

gallons of methoxide solution, I'll generate about 7.0 gallons of glycerin and about 35.7 gallons of unwashed biodiesel. To demonstrate this observation, I've included pictures of a sample collected from my processor immediately after it has been mixed for three hours at a temperature of 125 degrees Fahrenheit as well as after it has been allowed to settle overnight.

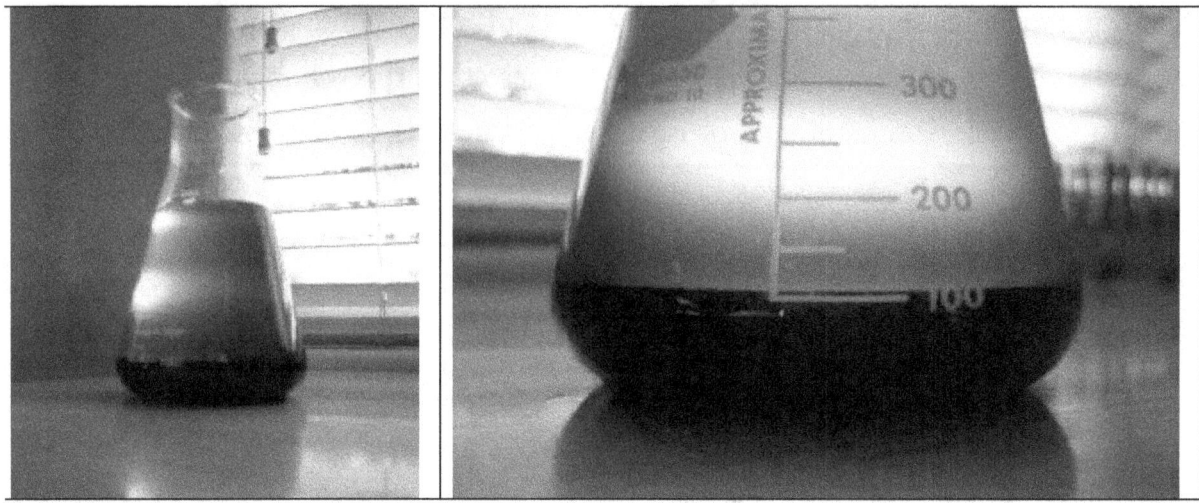

Photos of sample after it has been allowed to settle overnight.

500 ml of the mixture separated out into about 100 ml of glycerin and 400 ml of unwashed biodiesel. The percentage of glycerin in this sample (~100/500 or ~20%) is slightly less than the volumetric percentage of methoxide added to my processor (7.7/35 or 22%). Based on this sample, I expect to see about 7 gallons of glycerin collected from my processor.

Please note that my yields of unwashed biodiesel and glycerin will vary somewhat (+/- 1 gallon) from batch to batch. Typically, I'll generate about 7 gallons of glycerin for 35 gallons waste vegetable oil which has been base-transesterified with 7.7 gallons of methoxide.

It has been noted on some reputable blogs that yields of biodiesel are dependent on the kind of oil used for processing. I suspect that the quality of the oil you are using also has a significant impact; specifically, acidic or overused oils tend to increase yields of the "glycerin" phase at the expense of unwashed biodiesel. Also, note that final yield for finished (washed and dried) biodiesel will decrease during the washing process since soaps suspended in the biodiesel phase will result in biodiesel being lost with wash water.

5.3 Building an Appleseed Processor

The Appleseed processor is essentially a converted electric water heater used for the purpose of making biodiesel from waste vegetable oil.

The following is a summary of the steps and supplies used to build an Appleseed processor at El Instituto Tecnologico de Nogales (ITN) which is located in Sonora, Mexico. ITN is currently a partner in an Environmental Protection Agency (EPA) Border 2012 Project known as "The Ambos Nogales Biodiesel Capacity Building Project". The goal of the project is to create economic incentives for the recycling of waste vegetable oil and grease for the purpose of keeping this material from being disposed of improperly

With respect to the project, ITN is not funded to actually make biodiesel. Instead, it is tasked with evaluating potential oil sources in Nogales, Sonora, and developing a laboratory where samples rendered from other project stakeholders could be evaluated for quality prior to submission for ASTM testing. However, through a generous donation provided by a non-profit known as Friends of the Santa Cruz River, ITN was able to supplement their laboratory with an Appleseed processor and wash tank. This will ensure that students not only learn how to evaluate the quality of biodiesel, but also make and document the process of rendering biodiesel from waste vegetable oil. It's hoped that the Spanish documentation generated by ITN will be shared over the web so that other (Mexican) municipalities that are challenged by improper disposal of waste vegetable oil can learn how to turn this problem into an opportunity for offsetting fuel costs.

I took this opportunity to supplement the diagrams and processes documented in Maria Mark Alovert's manual. I made some slight modifications to Maria Mark's suggested design for the processor, as a result of having some extra plumbing supplies and ball valves; As such, some of the components may be considered optional.

Diagrammed Processor

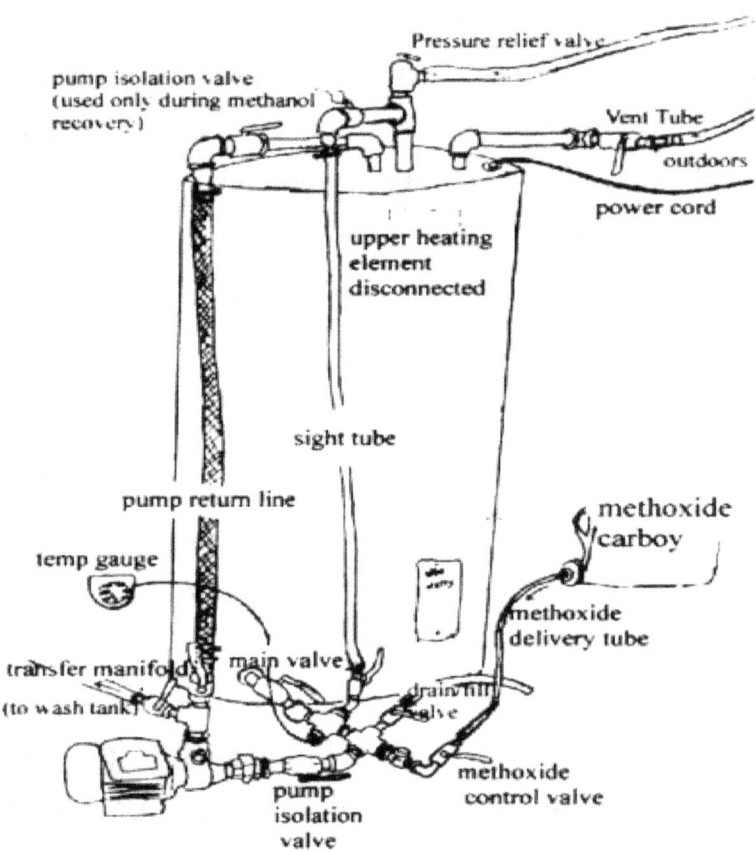

Short circuiting the upper element of your water heater

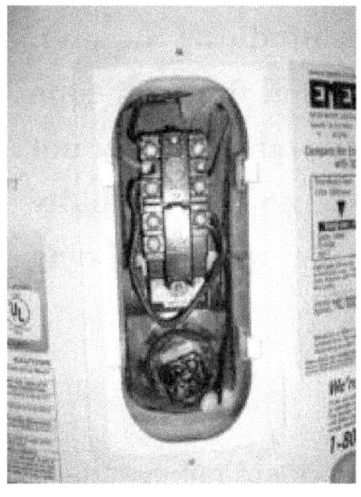

Upper element wiring

As noted by Graydon Blair (Utah Biodiesel Supply): "Notice that I've removed the wires going to the top element and by-passed the top element's thermostat. Also notice that I DID NOT by-pass the upper limit temperature control switch, but went through it. This is on purpose for safety."

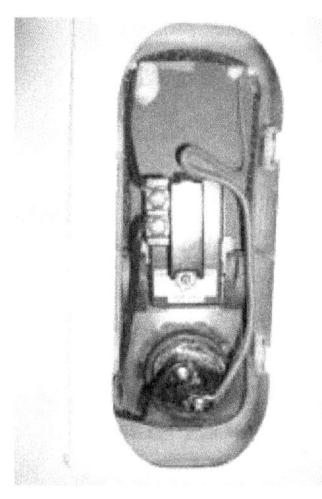

Lower element wiring

As noted by Graydon Blair (Utah Biodiesel Supply): "If you look closely you'll see that I've set the temp. Gauge to just a hair past the 125 deg F mark. This is where I start them out. I've found sometimes that "just a hair" is actually right at the 130 Deg. F mark."

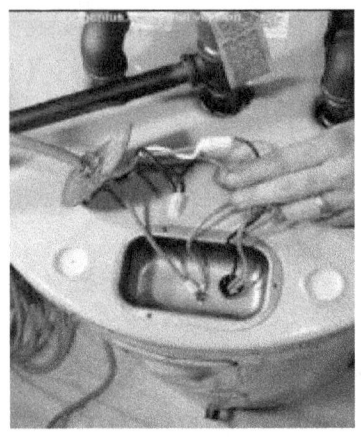

Junction box wiring

As noted by Graydon Blair (Utah Biodiesel Supply): "If you don't feel comfortable wiring, find someone that does. It's not worth shocking yourself from not wiring these things right. I assume no responsibility for your wiring errors. These pictures are being posted here for information purposes only."

Use a coat hanger to remove the plastic insert from the cold water port once nipples are extracted.

Remove plastic insert from one of the galvanized nipples; this nipple can be recycled for use as a vent.

Remove the galvanized nipples from the water heater using the ¾" internal pipe wrench.

Most anode nuts require a 1 1/16 inch socket (or 27 mm) to remove.

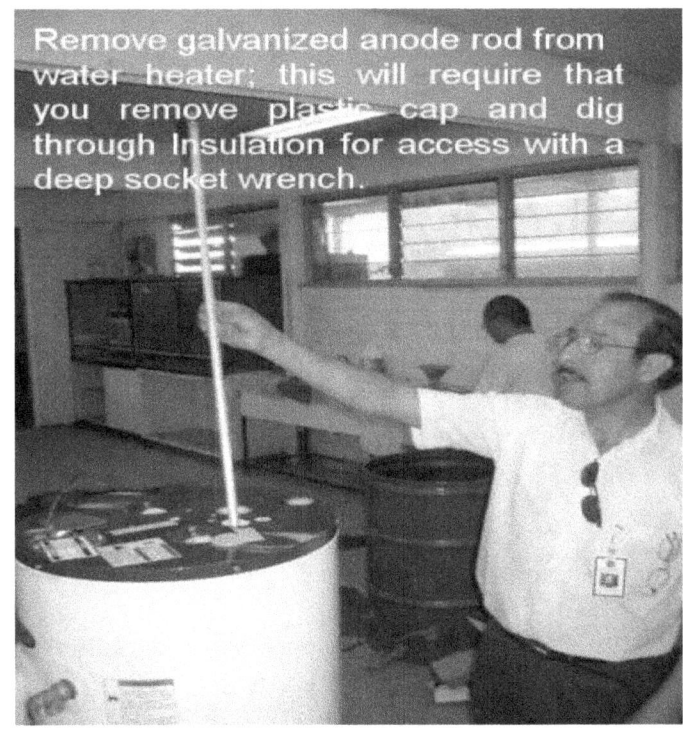

Remove galvanized anode rod from water heater; this will require that you remove plastic cap and dig through insulation for access with a deep socket wrench.

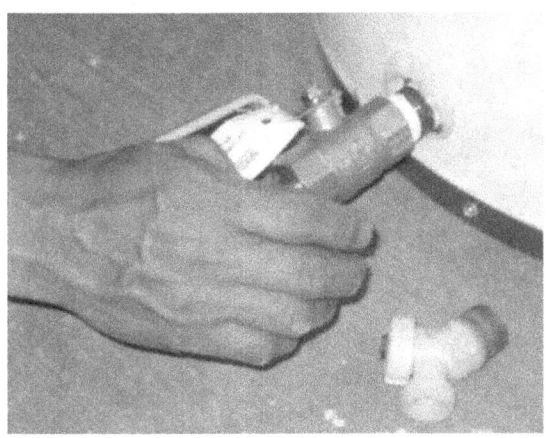

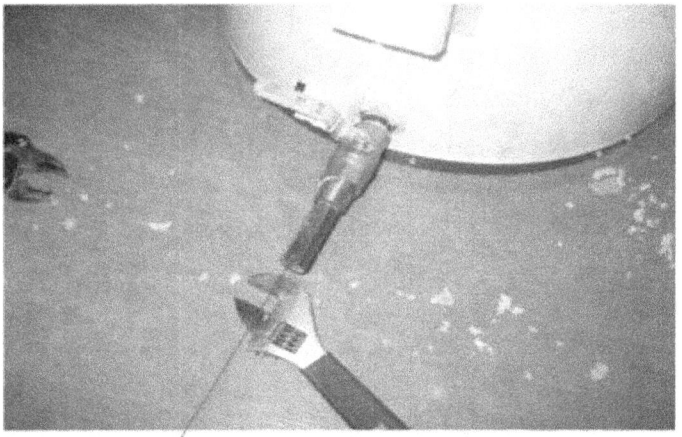

Use a ¾" internal pipe wrench and a crescent wrench to attach close nipples throughout construction; this will preserve threads on the nipples.

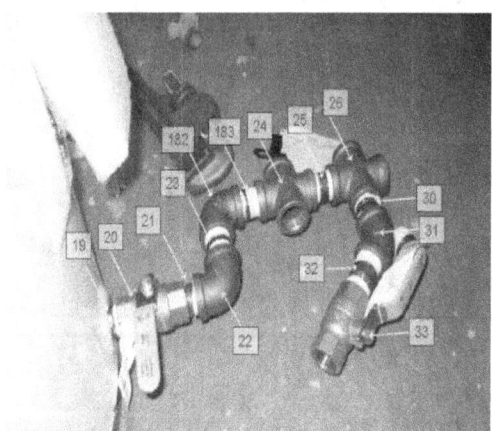

This component varies from Maria Mark Alovert's design;
Use to elbows so that you can vary the elevation of the pu
so that the pump won't hang;

(There may be other variations to Girl Mark's design in th
presentation based on materials available during design).

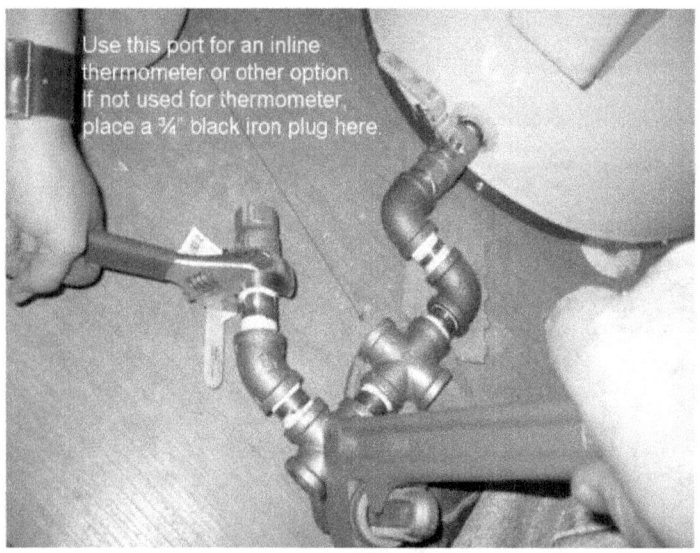

The 1 inch Clearwater pumps you purchase from Harbor Freight will require an extension cord. This shows how the extension cord should be wired. Note that the ground screw (green wire) is under the base of the black plastic enclosure; you will need to remove that base in order to get access to that screw in order to ground the green wire from your extension cord.

Your pump comes with a small piece of plastic that is used to clamp the extension cord wires in place. Don't forget to clamp the extension cord in place before sealing things up.

Finished wiring ready to be sealed with plastic cover shown on the right of the photo.

Note that this section has a ¾ X ½ " bushing with a ½" elbow and a ½ "ball valve.

Port to be used for an optional In-line thermometer; cap this port if thermometer is not used here.

Put the 1 X ¾ inch bushing on the manifold first before attaching the pump via the union.

Once you've finished wiring everything up, test the pump by plugging it into a ground faulted interrupt socket. You only need to test it for one second to know everything is wired properly. If the GFI "pops", you know you have a problem with your wiring.

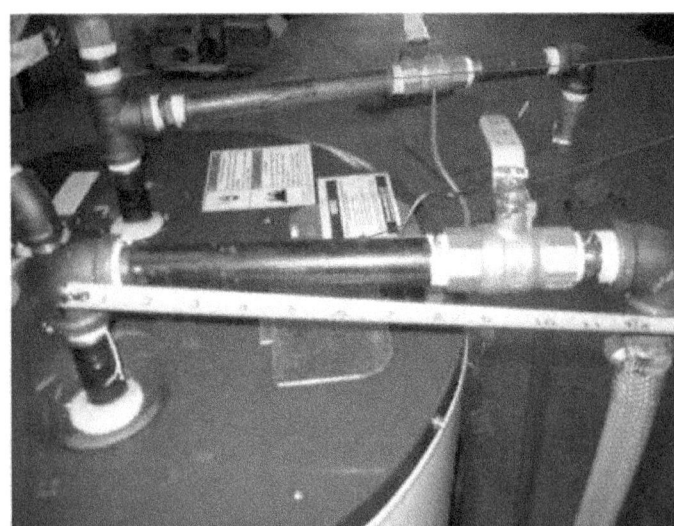

½ X 8" nipple

¾" X 8 inch nipple

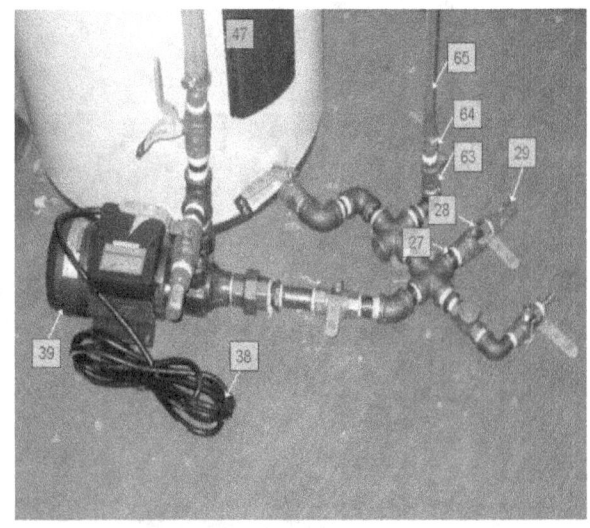

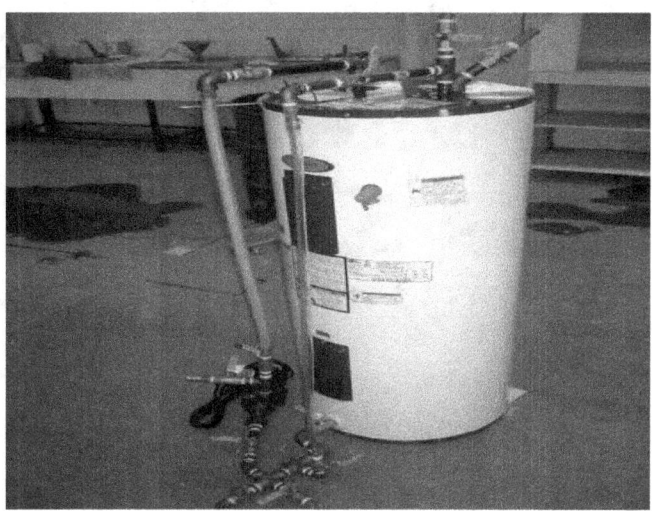

NOTE: use two clamps vs. single clamp shown for sight tube and return flow tubes on each Barb; do this on both ends.

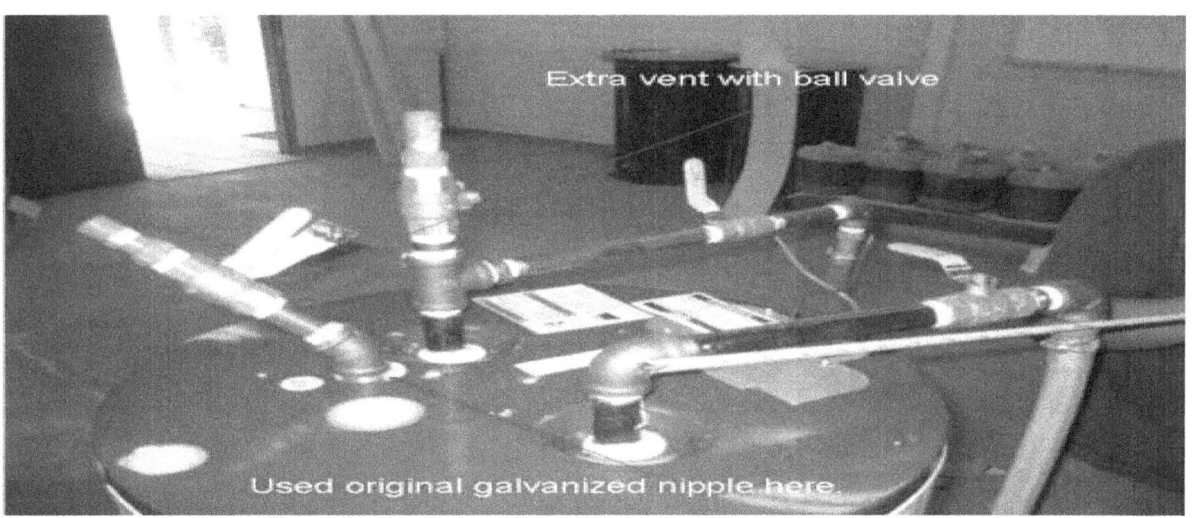

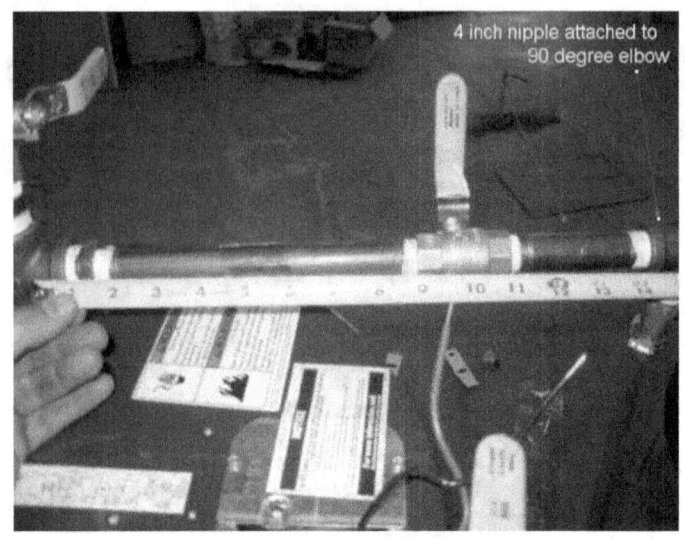

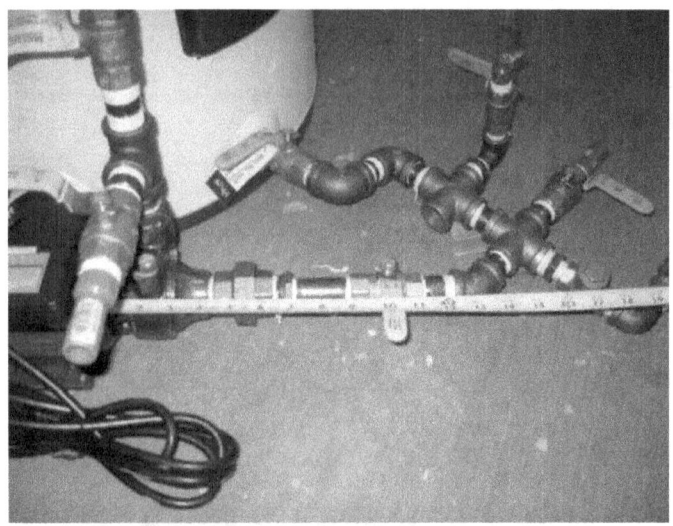

112

5.3.1 Diagrammed Wash tank

Here is a model of a completed standpipe washtank. You can also use these to store your oil. The standpipe helps separate wet oil from dry oil.

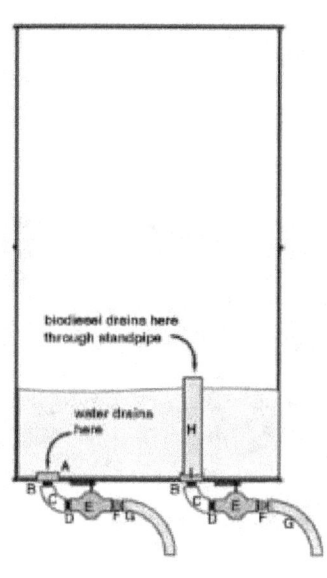

WASH TANK

Purchase five 2 X 4 X 8's and cut each into three 24 inch lengths and one 14 inch length

14 inches

All other pieces are 24 inches

¾" X 3 inch nipple

2 inch X ¾" bushing

¾" X 4 inch nipple

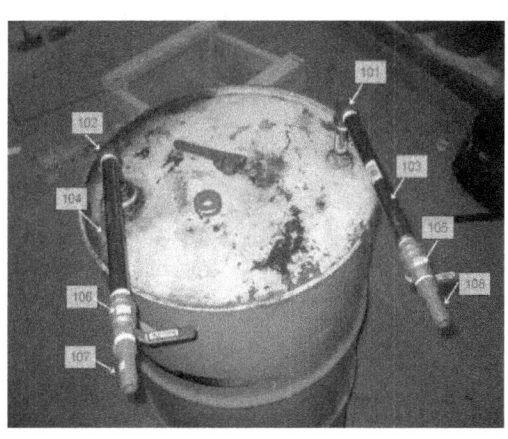

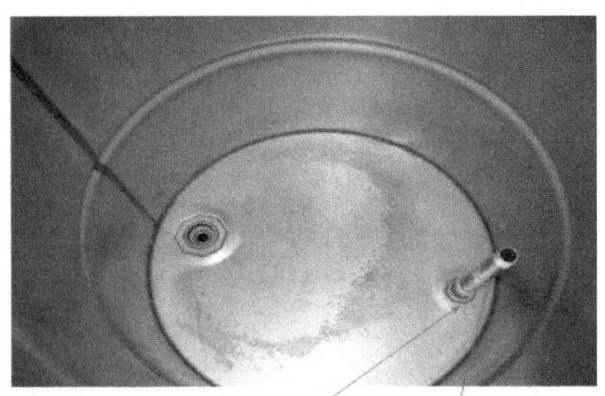

Add ¾ inch coupler to fasten 10 inch nipple (standpipe)

Add intermediate boards once processor is placed on stand.

Attach brackets to the stand.

5.3.2 Simplified Appleseed Processor

This section summarizes the processor I'm currently using to render biodiesel. This design is much simpler than the one presented in the prior section, but just as effective. The components for this design were purchased as a kit from on online distributor known as Biodieselwarehouse (142Hhttp://www.biodieselwarehouse.com) a few years ago. I do not have a parts list developed at this time; a review of the photos should provide enough guidance for reconstructing without having to purchase a kit. However, if you don't have any local hardware or plumbing stores, purchasing a kit isn't a bad idea.

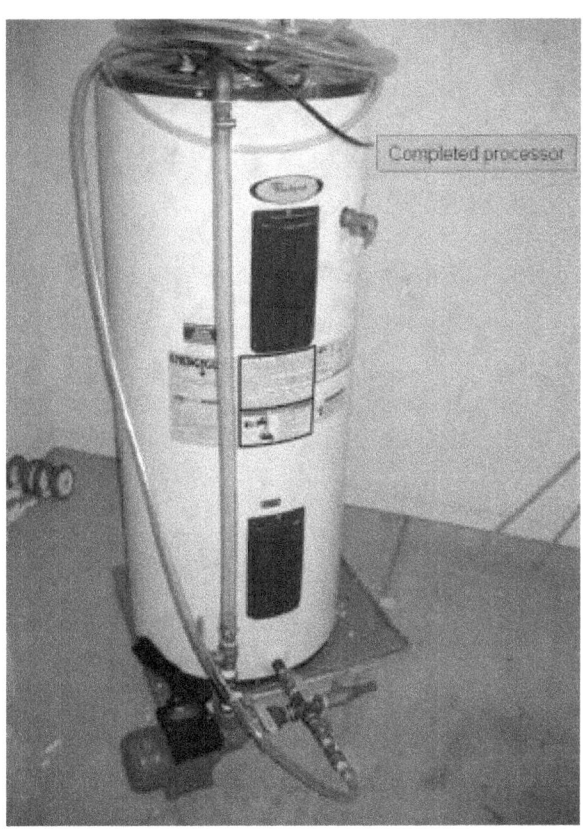

5.3.3 Waste-Oil Funnel

Most Appleseed processor kits will provide you with tubing for the inlet port on the processor pump, but no guidance on how to deliver the oil into the processor. One option is to build a second standpipe tank for waste oil storage, and then plumb the port attached to the standpipe to the inlet port on your pump. This is the method I'm currently using since it ensures I will deliver dry oil to my processor via the standpipe.

A second option is to build a simple funnel out of a bucket. Here, I've outlined my design for a simple funnel which can be used with standard 5 gallons carboys used to collect your oil.

Above is a photo of the completed funnel. The funnel is just a modified five-gallon bucket which has a bung at its base so that plumbing can be attached. The lid on the bucket is useful for keeping dust and other contaminants from entering your funnel when not in use. This bucket was originally used to store chlorine tablets for pool disinfection. Any bucket will do, but I recommend one that is five-gallons or greater in capacity.

Once completed, the funnel is placed on a couple of bricks to provide room for the plumbing at the base of the funnel. As an alternative, you can build a box/brace constructed from 2X4s. A brace to house the funnel is helpful since it provides a surface on which to lean your 5 gallon carboys while they are draining into the funnel.

The next photo shows the parts I purchased at my local ACE Hardware store for converting my bucket into a funnel. The part description for the bung is a WATTS PL-1842 ¾" Poly Body/Buna Washer Union. I also purchased an extra rubber washer to ensure a tight sealbetween the bung and the bucket, and some plumbing fixtures pieced together to create a seal between my bucket and the inlet port on my pump. Lastly, I like to use paint strainer bags (for paint sprayers) to filter for my oil. The bags fit perfectly in the opening of the bucket, and do a good job of keeping breading out of my processor.

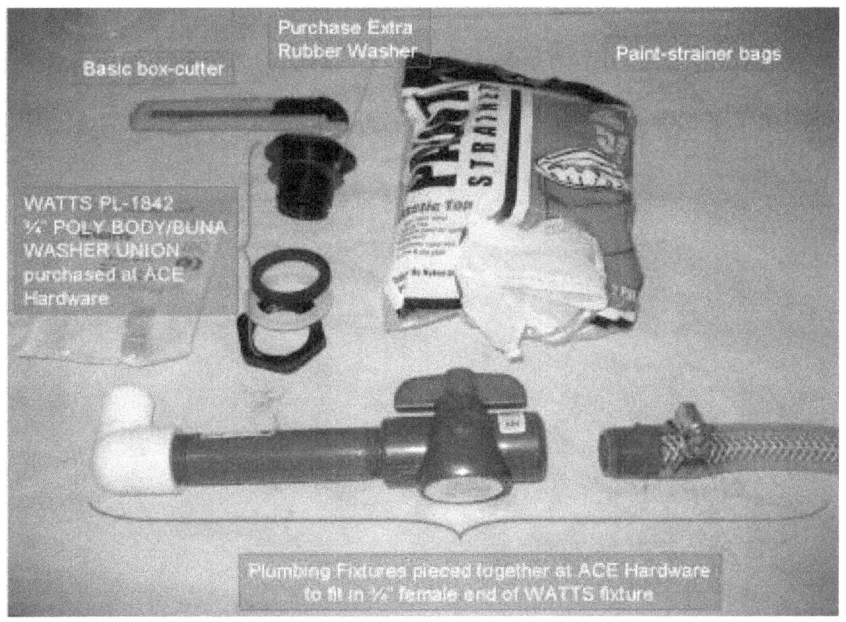

This next photo shows the installed bung at the base of my bucket. I used the box cutter to cut a hole in the base of my bucket to accommodate the bung. The hole was cut to match the diameter of the bung as closely as possible so as to avoid leaks. Note how the pieces that came with the bung kit are ordered to ensure a tight seal. The threaded part of the bung starts on the inside of the bucket and exits at the base and outside of the bucket. This approach will minimize pooling oil inside the bucket.

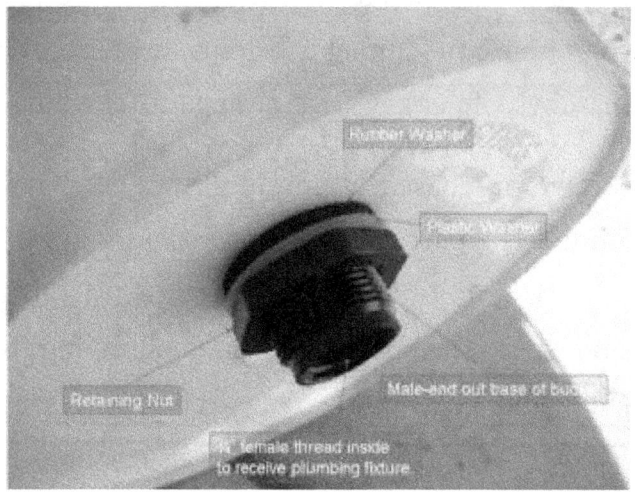

This next photo shows what the inside of my bucket looks like. The bung provides an exit for the oil at the base of the bucket. For filtration, I hang the paint strainer inside the funnel

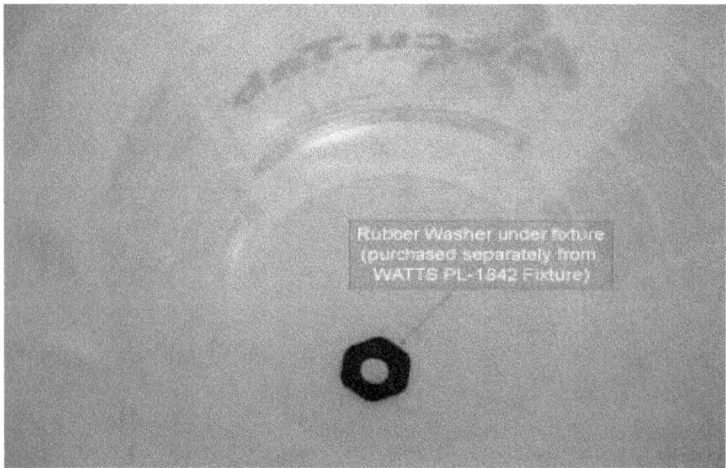

Below is a photo of the completed funnel turned upside-down. The bung at the base of the bucket contains a female thread which can be used to attach additional plumbing fixtures.

If you decide to use a funnel rather than a standpipe tank for delivery, let your collected oil sit in the sun before pouring it into the funnel. If you are using five-gallon carboys, the wetter/heavier oil will settle to the bottom of your carboys. As you are pouring the oil out of your carboys into the funnel, you will notice a change in the color and transparency as you hit that wet oil layer. At that point, stop pouring it into your funnel, and pour it into a separate carboy in the sun. Use this carboy as your receptacle for the dregs from other carboys. Again, the sun will help you segregate wet oil from dry, and you can repeat the process over and over again until what you have left over is unusable. That oil can be placed in a milk jug and disposed of in the garbage.

6 Making your own Biodiesel

6.0 Introduction

Biodiesel is an excellent alternative fuel for diesel engines. It is made from agricultural products grown within the Commonwealth and can be used by farmers. It is most commonly made from oil extracted from soybeans, and there is a lot of interest in biodiesel production. In general, biodiesel can be produced from any of the following: **pure vegetable oil (soybean, canola, sunflowers); rendered animal fats; or waste cooking oil.** The oil is converted to biodiesel through a chemical process called transesterification. Glycerin is removed as a byproduct of the reaction, and the resulting fuel can be blended with petroleum diesel, or used directly as a neat fuel. Biodiesel should be evaluated according to the protocols outlined in the Biodiesel Standard ASTM (American Society for Testing and Materials) D6751 before use. The basics of biodiesel fuel are discussed as well as the myths and questions about biodiesel usage. This new publication presents the procedures for producing biodiesel, with particular emphasis on small-scale production.

Many readers will not want to invest the time for produce your own biodiesel. These readers will still find this publication of value, because it explains the relatively simple procedures to make the products. In that sense, it takes some of the "mystery" out of an important fuel that we used directly, or indirectly, every day. Keep in mind that chemicals discussed in this report can cause injury. Do not attempt any of the described procedures until you are confident you understand the safety procedures. It is relatively easy to produce a product that is suitable for use in industrial burners or older diesel engines. However, it is more challenging to produce fuel that meets or exceeds the ASTM D6751 standards. Use of biodiesel that does not meet the ASTM criteria could result in engine damage and will undoubtedly void all manufacturers' warranties. If you are planning to use the fuel you produce in a modern diesel engine equipped with a high-pressure fuel injection system, you must pay particular attention to the amount of residual glycerin that is present as well as the water content of the fuel. If you are unable to evaluate your fuel sample according to the protocols described in ASTM D6751, or cannot afford to submit a sample for professional analysis, then you should not attempt to use the fuel you produce in modern agricultural machinery or industrial equipment. Also, if you are planning to use your home-made biodiesel for on-road applications, you MUST be fully aware of both State and Federal regulations regarding taxing and permitting procedures required for fuel distributors.

6.1 Glossary:

- **Triglyceride**: The major component of fat consists of three molecules of fatty acid combined with a molecule of the alcohol glycerol (Figure 1). Triglycerides make up a large portion of many types of lipids (fats).

- **Transesterification:** A reaction between a triglyceride and an alcohol in which the - $OCHCH_2CH_2.....CH_2CH_3$ of the triglyceride and the CH_3O- of the alcohol (methanol) trade places (Figure 2).

> **Methyl-ester**: A compound formed from an organic acid and a methanol-based alcohol. Titration: An analytical method used to determine total acidity (i.e., free fatty acids) of waste oil. A strong base (such as sodium hydroxide, NaOH) is added to waste oil in measured amounts. If an indicator chemical (such as phenolphthalein) has been added to a sample of the liquid being tested, then a color change will occur at the point when all available hydrogen ions in the acids have been neutralized by the base.

> **Esterification:** A condensation reaction through which carboxylic acids react with alcohols to form esters.

> **Acid value**: The amount of free acid present in waste oil as measured by the milligrams of potassium hydroxide (KOH) needed to neutralize it per gram of acid (unit: mg KOH/g).

6.2 Chemical reaction for biodiesel production – transesterification

Biodiesel is made through transesterification between triglyceride and alcohol (usually in the form of methanol). As shown in Figure 1, the triglyceride molecule is like a capital letter "E", where the three "arms' of the capital "E" represent three long-chain fatty acids. In transesterification, methanol molecules replace the "backbone" and link the "arms" of the capital "E" to form three linear molecules (Figure 2). This new linear molecule is called a "methyl-ester", which is the scientific term for biodiesel (Figure 2).

$$CH_2O-OCHCH_2CH_2\ldots\ldots CH_2CH_3$$
$$|$$
$$CHO-OCHCH_2CH_2\ldots\ldots CH_2CH_3$$
$$|$$
$$CH_2O-OCHCH_2CH_2\ldots\ldots CH_2CH_3$$

Figure 1. Molecular structure of triglyceride

$$CH_2O-OCHCH_2CH_2.........CH_2CH_3$$
$$|$$
$$CHO-OCHCH_2CH_2.........CH_2CH_3 \;+\; 3\,CH_3OH \;\xrightarrow{Catalyst}\; 3\,CH_3OOCHCH_2.....CH_2CH_3 \;+\; CHOH$$
$$|$$
$$CH_2O-OCHCH_2CH_2.........CH_2CH_3$$

| Triglyceride (oil or fat) | Methanol | Methyl Esters (Biodiesel) | Glycerol |
| (100 lb) | (10 lbs) | (100 lbs) | (10 lbs) |

Figure 2. Chemical reaction for biodiesel production - Transesterification

The theoretical ratio of methanol to triglyceride is 3:1; which corresponds to having one methanol molecule for each of the three hydrocarbon chains present in the triglyceride molecule, and is equivalent to approximately 12% methanol by volume. In practice, this ratio needs to be higher in order to drive the reaction towards a maximum biodiesel yield; 25% by volume is recommended. The catalyst can be alkalis, acids, or enzymes (e.g., lipase). The majority of biodiesel produced today is done with the alkalis-catalyzed reaction because this reaction

(1) requires only low temperature and pressure,

(2) Has a high conversion yield (98%) with minimal side reaction and a short reaction time,

(3) Is a direct conversion to biodiesel with no intermediate compounds, and

(4) Does not need elaborate construction materials.

6.3 Making Biodiesel

So... you want to brew some biodiesel.

Before proceeding, allow me make a suggestion: find out if there is a local coop or group of friends who are doing this and who might be able to provide you with the support you'll need to get you off the ground safely and responsibly. In the event you don't have that kind of support, I've detailed a cycle of my own homebrew operation for your review. The purpose of this summary is to demonstrate what it takes from start to finish producing a good batch of biodiesel using Maria Mark Alovert's "Appleseed processor" design. Please consider this an introduction. It is by no means is comprehensive of all the safety measures or considerations for making quality biodiesel.

6.3.1 Collecting oil

- ❖ First, you need to identify an oil source with reasonably good quality oil. Note that most restaurants are already served by commercial grease collection outfits, so be prepared to do at least as good a job as the commercial entity that is already serving the restaurant. Being a replacement to a commercial service provider will assure you a constant source of grease, but it also comes with responsibility of consistently meeting the needs of the restaurant. Those needs may include power washing the area where oil is stored.

It's more difficult to convince management to allow you to just occasionally collect from their bins with no formal commitments, but it can be done. I've had pretty good luck working with commercial chains as compared to mom-and-pop cafeterias towards this end, and this approach is probably okay if you are just starting out. Don't be shy about this approach; the worst that can happen is that management will say "no".

Given that I formally relied on an oil collection coop for oil, I'm relatively new to working with restaurants. For the time being, I am authorized to collect oil from three businesses with no strings attached. However, I'll be the first to admit that this is not a sustainable model. Pending purchase of appropriate equipment, and guidance from others in the homebrew community, I hope to be a regular service provider to a single restaurant in the near future.

It is my understanding that the owner will make dewatered, filtered, low acidity oil available to the homebrew community. This is a good way to cut your teeth on home brewing without getting into commitments with restaurants that you may not honor at a later date. Whatever you decide, please don't take oil without permission from management of restaurants; that is considered stealing and reflects poorly on the biodiesel/SVO community at large.

Oil collection trailer for servicing restaurants on a permanent basis

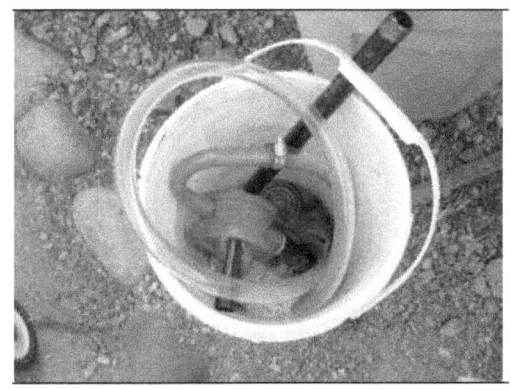

My alternative: a simple hand pump broken down and stored in a bucket for easy transport; simple, but effective.

- ❖ Secondly, for the time being, I can get by with Harbor Freight hand-pump and eight five gallon carboys. I use a seamless system consisting of flexible plastic tubing between my pump and my carboys. I use hose-barbs to hook it all up. This gets the grease into my carboys without spilling a drop. The system is very inexpensive and perfect for someone who is just starting out. Note that there are easier and more efficient approaches (i.e. trailers with 55 gallon drums; use of battery powered pumps, or providing restaurantswith carboys for oil collection). However, these are more expensive and can be postponed until you are prepared to provide service on a regular basis to a restaurant.

- ❖ When collecting oil, I never collect from a bin that is near empty, or from bottom of a bin since this is where contaminants settle. Instead, I place the suction portion of my pump just under the surface of the oil. I use a modified 2X4 length of pine and the hardware included with the hand pump to create an adjustable brace. This allows me to anchor my pump and set the inlet depth as needed.

This is a setup for collecting oil from a local restaurant. Note the cardboard box under the carboy to ensure I don't leave any spills behind. You can typically find cardboard in dumpsters located near the grease bins.

This seamless system prevents spillage during collection. Make sure to open the vents your containers before filling with oil.

Oil migrated to the back of car. I've placed a garbage bag at the base of my trunk to capture any oil drips. This demonstrates you don't need a truck to collect oil.

6.3.2 Filtering and settling the oil

- ❖ Once collected, I let the oil sit in the carboys for a day or two in the sun. This ensures that any breading or water captured during the transfer to my carboys settles and can be separated out for disposal.

This oil is settling in a yard well out of site from neighbors. Try to minimize the visibility of your operation so as to avoid neighbors filing complaints with the city or county

❖ Next, I strain the oil from the carboys through a 200 micron mesh for storage in a 55- gallon standpipe drum. I try to do this on a warm day since tallows in cold grease have a tendency to clog the mesh. This requires intervention on my part via gentle scraping with a spatula.

Since this picture was taken, I've placed a large piece of cheap plywood next to the

house. This keeps oil splatters from staining my stucco.

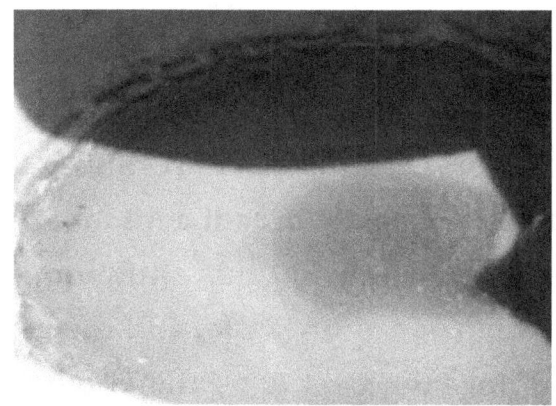

This is what your 200 micron filter will look like after filtering 5 gallons of heavily breaded oil that hasn't been allowed to settle and separate. Also, try to filter oil on warm days; the heat lowers the viscosity of the oil and keeps tallow from clogging your mesh. This makes filtering much easier.

- While pouring oil onto the filter mesh, I may notice a color change near the end of the pour. That darker oil is wet and contains most of the breading. At this point – I'll transfer the remnants (about half gallon) to a separate bucket. That bucket will receive multiple pours from others carboys, and will also be allowed to sit in the sun. I'll then repeat the process of pouring and separating good oil from wet oil. Any unusable leftovers are non-hazardous and may be disposed of safely in the garbage in a sealed container.

Place wet and/or heavily breaded oil in a separate container; avoid introducing this material into your standpipe barrel. Note the plastic barrel lid on the ground. I use this to capture drips from delivery tube of storage tank. This is also a good workspace for filling your carboys with oil while avoiding oil stains on the ground. The vinyl tubing you see was eventually replaced with a PVC elbow.

❖ Eventually, I'll remove accumulated breading from my filter mesh. This breading may also be disposed of the garbage in a sealed bag or bucket. (Milk jugs are also excellent for cleanly disposing of unwanted materials).

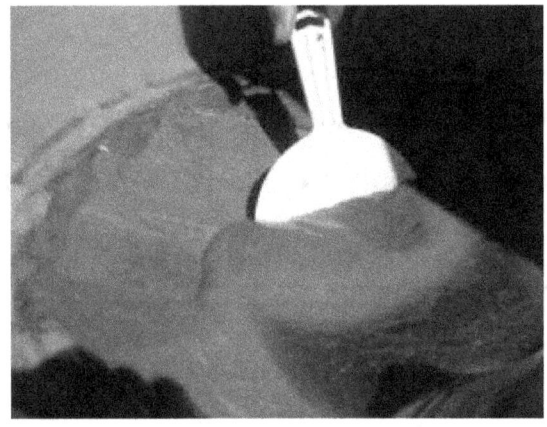

This is a worst case scenario resulting from filtering heavily breaded oil on a cold day. You can avoid this situation by allowing your oil to settle, and then pouring only the first 4.5 gallons out of a 5 gallon carboy.

❖ I may let the oil in the standpipe tank settle for a few days, preferably in the sun. Again, this allows any additional fine breading and/or wet oil not captured by my initial separation to settle to the bottom of the barrel. Wet and/or breaded oil is drained through the non-standpipe drain on your storage tank.

Filtered and settled oil ready for processing.

6.3.3 Migrating the oil to the processor

- ❖ Next, I open the standpipe drain in my barrel in order to collect oil for processing. Using the standpipe ensures that the oil I'm about to put into my processor is free of any debris and/or water that may have been introduced from my carboys and settled out in the barrel. The standpipe doesn't need to be any taller than 4 inches; since most of the breading will be captured by the strainer.

This is waste vegetable oil on a cold day; notice lack of transparency due to cold weather clouding.

This is the same sample on a warm day in the sun; notice how the heat helps clear up the oil.

- ❖ I'll fill a five-gallon carboy with filtered oil and migrate to my Appleseed processor seven times (35 gallons total). To do so, I'll pour the filtered/dewater oil from my carboys into a large funnel made out of a bucket and stand. The stand gives me something to rest the carboys while draining, and the funnel provides me with a clean way of getting in the oil into the processor.

 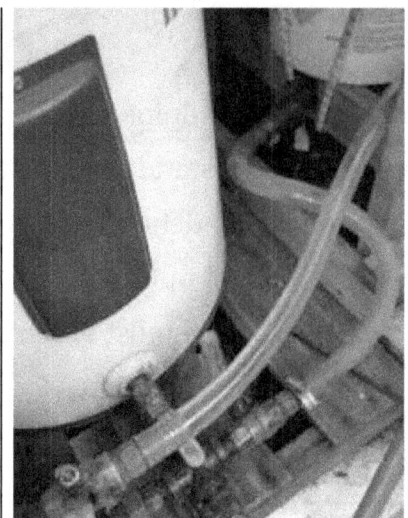

Above are photos of my bucket funnel in a stand, and a close up showing how the bucket is attached to the suction port on my processor. The pump in front of the bucket is part of my wash water disposal system.

This shows how my ball valves are set when I'm delivering oil to the processor via my funnel on the right. As an alternative, I can attach the suction port directly to my oil barrel. Note: The line containing biodiesel (at the base of the picture) is from my last batch. I place ball valves at the end of my lines to help keep things from getting too messy.

Here, I'm pouring oil out of my 5 gallon carboys into a funnel made from a five gallon bucket. The funnel is attached to a suction port on my processor pump. Note that there's a paint filter-netting purchased from a local hardware store. This serves as a secondary filter for oil prior to introduction into my processor.

On cold days, the netting may clog with animal fats (tallows) which solidify at higher temperatures. Note that tallows will increase the gel point of your fuel.

- ❖ The funnel is attached to a suction port on my processor's pump. Of course, I'll have the pump engaged in order to transfer the oil to my processor. Before engaging the pump, I make sure the ball valve controlling the return flow from the bottom of the tank is closed. I'll make sure my glycerin/biodiesel delivery ball valve located above the pump is closed. Even though there is a reverse-flow check valve on the methoxide delivery port, I also make sure that port is closed.
(As an alternative, I can also hook up my standpipe oil tank directly to my Appleseed, and transfer oil until the level reaches a 35-gallon mark on my sight

tube. This is much cleaner and easier for delivery, but it is a bit harder to get exactly 35 gallons into my tank. If the weight of the oil is too heavy to handle, use this alternative approach).

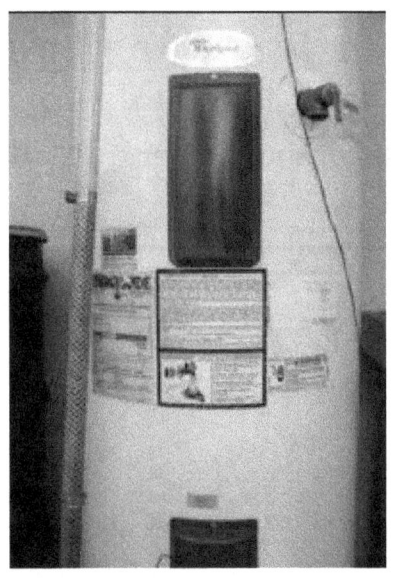

Here's my processor filled with 35 gallons of oil. Note the hose clamp on the sight tube. This helps me determine when I've delivered my desired volume in the event I'm not introducing oil through the use of measured carboys.

- Once 35 gallons have been transferred, I close the valve from my funnel, and open the return valve from the bottom of the processor. In a few seconds, I should be getting a steady flow of oil circulating throughout the Appleseed

The sight tube will be solid with oil up to the return assembly when the oil is mixing as it should. You'll hear the oil mixing if all is

working properly

- ❖ Once the oil is well mixed, I'll take a sample from the glycerin/biodiesel delivery port by slowly opening the ball valve and collecting the oil in a jar. I only need a few milliliters for an acid titration.

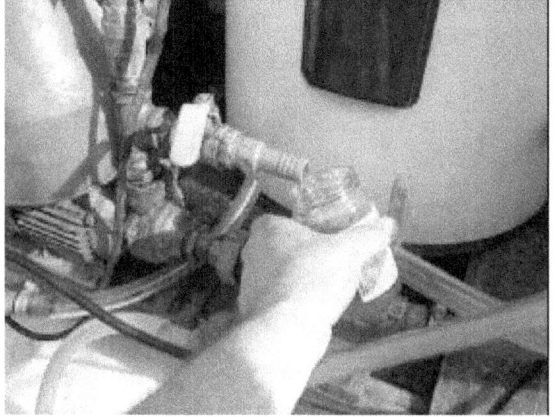

I'll let my oil mix through the processor for 10 minutes before collecting a sample. I only need to open the ball valve above the pump slightly to collect my oil.

6.3.4 Evaluating the quality of the mixed oil

- ❖ Conducting an acid titration on your oil each and every time I make biodiesel is important for determining the amount of catalyst necessary for my methoxide recipe. The titration helps me determine how much extra catalyst is needed to neutralize the free fatty acids (FFA) that may be present in my oil. This is important since I don't want to short the mass of catalyst needed for transesterification- something that occurs through the consumption of catalyst in FFA neutralization.

Being as accurate as possible is important to minimize the amount of washing required later in the process. I want to make sure I add enough catalyst to ensure free fatty acid neutralization and complete transesterification. If I don't add enough catalyst, my biodiesel will contain unreacted mono and diglycerides. If I add too much, the extra catalyst will need to be washed out during the biodiesel washing phase.

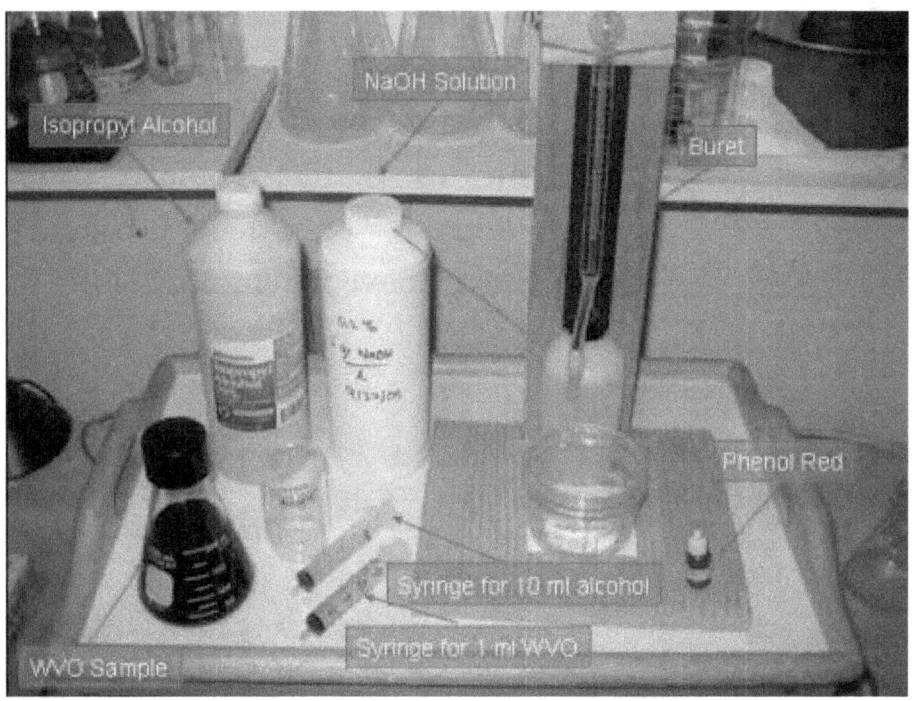

This is a simple kit for determining the acidity of our waste vegetable oil. It contains baby medicine syringes which you can get for free from your local pharmacy; isopropyl alcohol (purity above 90% is best, but 70% will work if that's all you can get); phenol red indicator; a 10 ml or greater burette for titrating our waste vegetable oil with the catalyst solution; and a 0.10% catalyst solution. For my catalyst solution, I add 1.0 gr of catalyst per liter (1000 ml) of distilled water to get the desired concentration. This solution gets added to your burette as the titrant. In this case, my catalyst is sodium hydroxide (NaOH).

- ❖ For the titration, I take 10 ml of fresh isopropyl alcohol purchased from my local drugstore, and add it to a baby food jar. I'll then mix the alcohol with 1 ml of oil and a drop or two of phenol red. I then swirl the solution to ensure that my oil dissolves in the alcohol as thoroughly as possible. If there is acidity in the sample, the solution will turn yellow.

The solution of waste vegetable oil, phenol red, and alcohol gets stirred until it appears as a homogenous yellow mixture.

- ❖ Next, titrate the mixture with the 0.10 % catalyst solution until I see a color change from yellow to red. The catalyst can either be sodium hydroxide, or potassium hydroxide. By "titrate", I mean adding my catalyst to the oil solution one drop at a time from the burette. As I'm adding the catalyst, I swirl the solution gently ensuring that I don't spill or splatter the mix. While mixing, the color change should hold for at least 15 seconds (preferably 30 seconds) in order to be considered complete. The milliliters of solution added to the oil for the color change equates to the acid number of the oil sample.

Slowly titrate the solution with the 0.1% solution of catalyst delivered from the burette. The solution is swirled during the titration. Eventually, the color of the solution will change from yellow to pink. The color change should hold for 30 seconds while the solution is being

swirled. When this happens, I'll know all the FFAs in the WVO have been neutralized with the titrant.

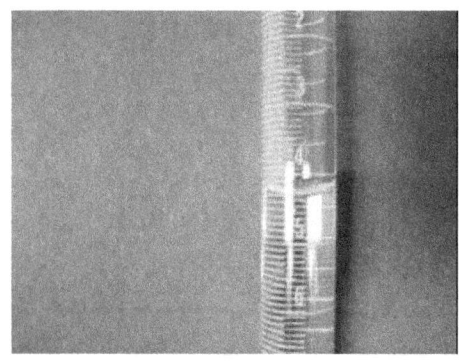

Initial reading on titrant: 4.3 ml

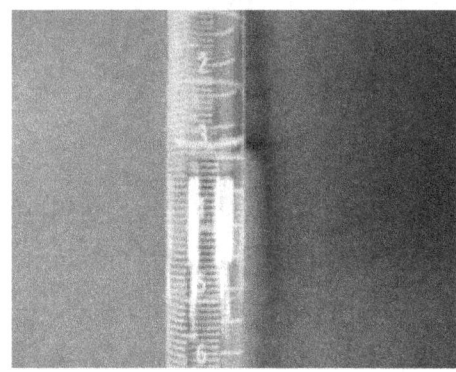

Final reading on titrant: 3 ml

The amount of titrant I added is calculated by subtracting the final reading from what I started with. In this case, 4.3 ml – 3 ml = 1.3 ml

(NOTE: You may use potassium-hydroxide or sodium-hydroxide for your catalyst. Potassium hydroxide is much easier to dissolve in methanol. I personally use sodium hydroxide because I can make hard soap from the residual glycerin, whereas potassium hydroxide will produce soft soap. Whichever you decide to use, make sure you prepare your 0.1% titration solution with the same catalyst you are going to add to your methanol.)

6.4.5 Determine the recipe for the methoxide

- ❖ The amount of catalyst I added during the titration will be used to calculate how much catalyst I must added to my methanol. I use a simple recipe calculator available at 148Hhttp://www.biodieselcommunity.orgto determine the total amount of catalyst required.

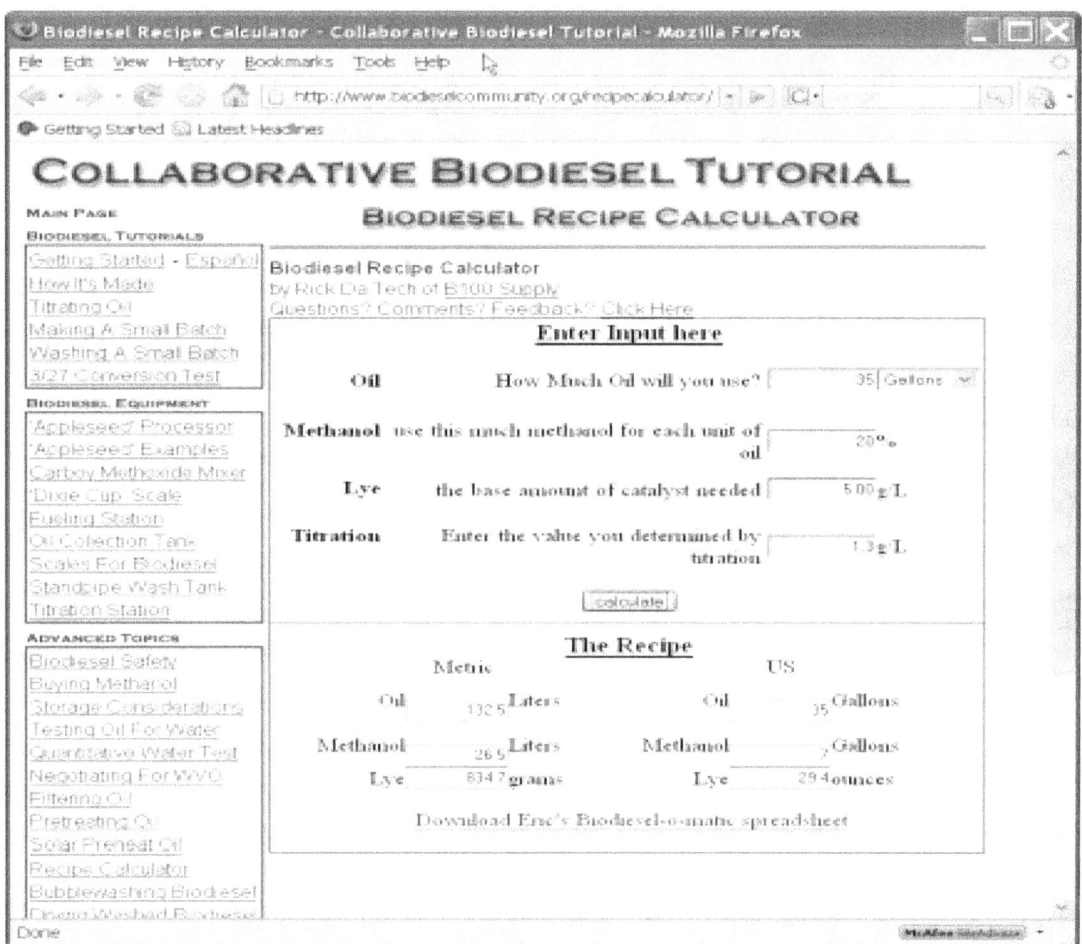

For good quality methanol, 20% should suffice. However, if methanol is potentially contaminated with water, increase the percentage to at least 22%. If you are uncertain, just use 22%.

- ❖ For a 35 gallon batch of oil, I'll typically use 7 gallons of methanol split between two 5 gallon carboys (3.5 gallons per carboy). I'll split the required amount of catalyst between two containers so that it can be distributed equally between my two carboys for preparation of methoxide. This will require weighing the calculated amount of catalyst, and splitting the mass between two containers. Of course, I'll be wearing safety goggles, a long sleeved shirt, and gloves to avoid skin and eye contact with the catalyst.

(If you read the MSDS for sodium hydroxide or potassium hydroxide (catalyst), you'll understand why you don't want to get this material on your skin or your eyes. MSDS for respective chemicals are in the appendix.)

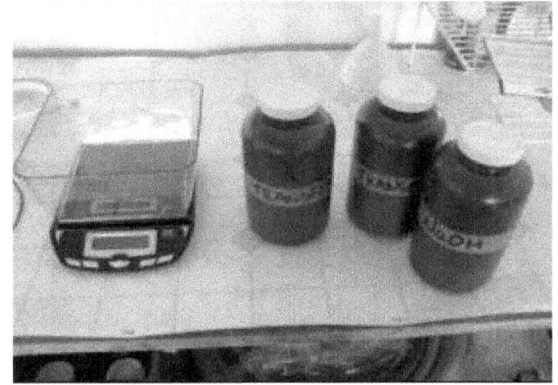

Here is my scale (accurate to 1.0 gram), two transfer containers for my catalyst, and a container of sodium hydroxide (NaOH).

If you purchase sodium hydroxide, you get the "beads" rather than the "flakes". Beads have a higher surface area, are much easier to dissolve in methanol, and are easier to transfer via a funnel.

Use a glass rod or a wood pencil to help get any clumps of sodium hydroxide through your funnel and into your transfer container. Never use your bare hands.

According to Industrial Safety Supply (149Hwww.lss.com), Ansell "Touch-n-Tuff" nitrile gloves are chemically resistant to sodium hydroxide, sulfuric acid, and methanol.

If you purchase disposable nitrile gloves, remember to use these once and throw them away since they will degrade with exposure.

More information on gloves and chemical resistance is presented in the appendix of this document, and may also be researched at the website hosted by Industrial Safety Supply (150Hwww.lss.com). For your own protection, do your own researches before you purchase?

Here is one transfer container filled with 417 grams of catalyst and ready to go. Note that prior to weighing, I zeroed my scale to account for the weight of my funnel and container.

6.4.6 Warm the oil

❖ My processor has a manual interrupter to complete the circuit that delivers 220V AC power to a single 2000 watt heating element at its base. Of course, this is running through a 25 amp circuit breaker, and all my wiring is of appropriate gauge and encased in conduit. Once my oil is circulating smoothly through the processor, I'll turn this switch on so that the oil can begin heating.

(I could have also replaced my 220V 2000 watt element with 1500 watt element rated to run on 110V circuit. This would allow me to plug in my processor to a grounded wall outlet. Note that heating times for oil will take longer. Also, the appropriate gauge extension cord must be purchased to power the element).

Switch for water heater.

Wiring for my water heater is encased in conduit.

❖ I have a cheap Harbor Freight electronic indoor/outdoor thermometer with the window sensor attached to my processor tank just above the bottom water heater element. I use the "outdoor" setting on the thermometer to measure the temperature of the oil in my processor. I'm aiming to get the temperature of my oil between 125 to 130 degrees Fahrenheit. With my initial oil temperature at 65 degrees Fahrenheit, this takes approximately 30 minutes.

(NOTE: Avoid adding methoxide to oil that is heated above 135 degrees Fahrenheit. Although the boiling point for methanol is 148 degrees, you'll want to take into consideration the inaccuracy of your thermometer).

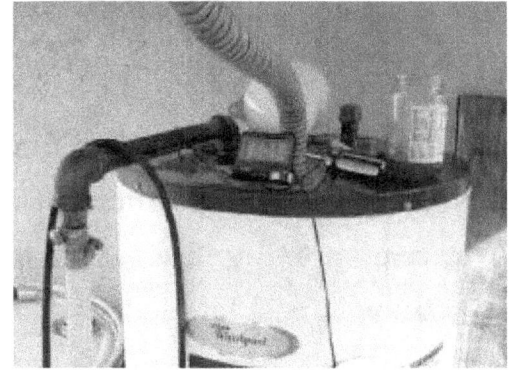

Here is a cheap harbor freight indoor/outdoor thermometer. The setting on the thermometer is reading "outdoor". Since the sensor is attached to my water heater under the insulation, the "outdoor" setting is really measuring the "indoor/inside" temperature of the oil in my processor.

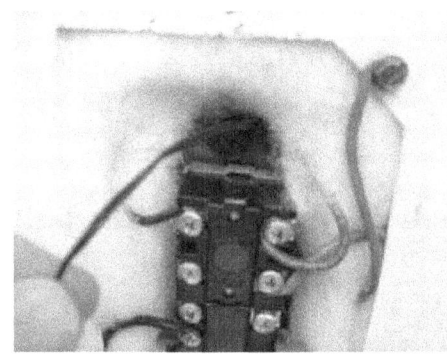

Thermometer window sensor attached to water heater.

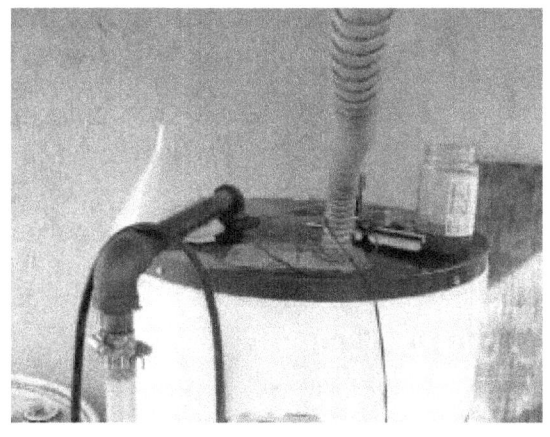

I keep the thermometer under a funnel to protect it from the sun when not in use.

6.4.7 Make the methoxide

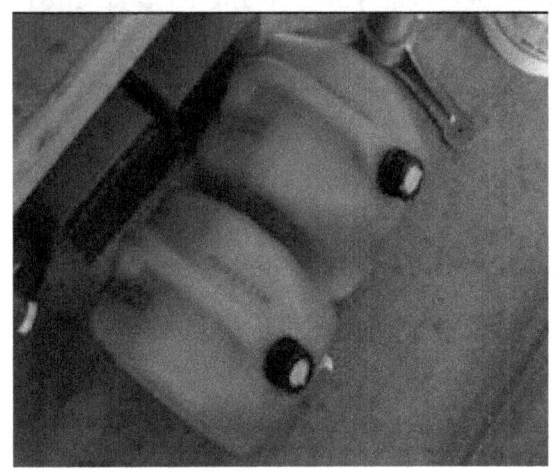

For 35 gallons of vegetable oil, my recipe calls for 7 gallons of methanol split between two 5 gallon carboys (3.5 gallons per carboy).

When purchasing methanol by the gallon, I use appropriate containers approved for transporting fuel. I also make sure they are labeled with respective contents. I always purchase my methanol the day I am going to use it, and I avoid storing methanol on site in any quantities.

I could save some money by purchasing methanol in 55 gallon barrels, but the associated risks associated with storage are not worth the cost savings.

Here is my carboy with a ball valve attached. The ball valve is currently open.

Note the 3.5 gallon mark on the carboy. This helps me better determine the point at which to stop adding methanol to my carboy from the red container. Use a fuel funnel for respective transfers.

Here are my two carboys ready to receive measured amounts of catalyst as determined from my recipe.

- While the oil is warming, I'll add my catalyst to the carboys containing the methanol. I always wear safety goggles and gloves when mixing these chemicals, and I keep upwind of my carboys when adding catalyst. I also keep running water nearby so that I can wash any chemicals that my skin may come into contact with. Most important of all, I NEVER look directly into the carboy opening while adding my sodium hydroxide since there's always a possibility that some methanol might splash out of the container and onto my face.

(NOTE: If you read the MSDS for methanol as well as sodium or potassium methoxide, you'll understand why you don't want to get these materials on your skin and especially your eyes. MSDS for respective chemicals are in the Appendix.)

Here, I am pouring the catalyst from my transfer container into my carboy. I'll stand upwind while pouring, and will be wearing a long sleeve shirt, long pants, gloves, and goggles in order to protect myself from any splashes.

- Once the catalyst is added to the carboys, I'll seal the vent and the cap, and being mixing the contents of the carboys by swirling for at least 30 seconds. Once mixed, I'll set the carboy down and gently open the breather valve on the carboy. The reaction of methanol with the catalyst is exothermic, so you may feel the carboys get a little warm while mixing. It also causes some out gassing, so relieving pressure via the breather valve on the carboy is important.

I may have to mix the solution several times over the period of an hour to ensure that all the catalyst is dissolved in the oil. If it's not dissolved, I'll be able to see solid catalyst at the base of the carboy. I usually allow my catalyst to dissolve in the carboys while my oil is warming up in the processor.

Here, I'm mixing my catalyst and methanol. My lids and/or ball valve are tightly fastened and closed. I swirl the solution while crouching in order to save the stress on my back. The reaction is slightly exothermic, so it will release heat and will also cause the headspace to come under pressure.

After mixing for about 30 seconds, I'll set the carboy down and open the vent. I'll face the vent away from me when opening so as to avoid any contact with fluids under pressure.

- ❖ Any unused materials are clearly labeled and stored outdoors in a sealed Rubbermaid container. I make sure this container is inaccessible to children or animals.

I keep all my chemicals in this container and under the shade of a tree. Containers with fuel are properly vented to avoid buildup of pressure and potential rupturing.

Adding methoxide to the heated oil

- Once the oil has reached the 125 – 130 degrees, I'll cut power to the water heater element by removing my switch. This is very important; never add methanol or methoxide to a water heater that has a live element. You are risking a fire (or worse) if you do.

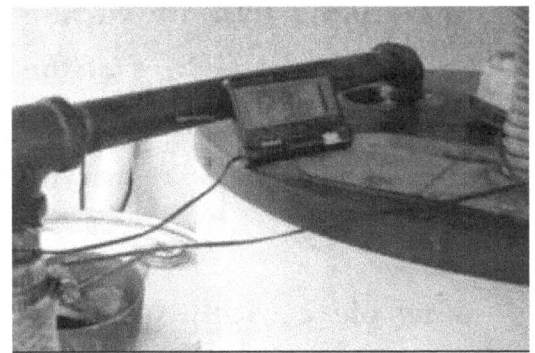

Temperature reads 129.6 F; I am now ready to cut the power off to the water heater.

- Next, I attach the carboy with the sodium methoxide to the delivery port on my processor. Then, I'll open the breather cap and the ball valve on the methoxide carboy, and then slowly close the ball valve associated with the return flow of oil from the bottom of the processor. This will generate a negative pressure that will start to suck the methoxide out of the carboy. As soon as I see a color change in my sight-tube, I'll stop closing the ball valve at the base of the processor. The key here is to maximize the mixing of the sodium methoxide with the heated waste vegetable oil. As such, you don't want the methoxide to be introduced too quickly into the mix. It should take at least 5 minutes for each carboy to empty.

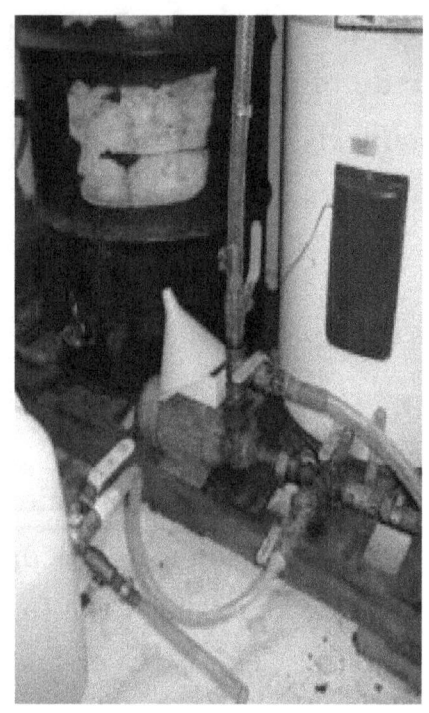

Note that there is a check valve associated with that port to prevent circulating oil from getting pumped into my carboy.

Here, the ball valve on the carboy and the ball valve on the manifold are both open. No methoxide will enter the system until I create a negative pressure by slowly closing the ball valve at the base of my processor.

Here, the ball valve at the base of my processor has been cut back. Note color change in my sight tube. Try to maximize mixing by only cutting back slightly on the ball valve at the base of the processor; the 3.5 gallons of methoxide should take at least five minutes to drain from the carboy.

The real mixing is taking place within the sight tube, not within the processor. As such, maximize this mixing as much as possible by introducing your methoxide slowly

Note the light color of the methoxide / vegetable oil mixture.

- ❖ When a methoxide carboy is near empty, I may have to tilt it to get the last bit of methoxide out of the container. As an alternative, I sometimes place a rock or brick under the carboy to help drain. I also make sure I return all the ball valves to their original positions when replacing an empty carboy with a full one.

Continue mixing the solution for 2 to 3 hours

❖ Once all the methoxide has been added (7 gallons), I'll return all the ball valves to their original position so that my mixture is being circulated throughout the processor. In order to assist the transesterification, two hours should suffice (some prefer three hours). I have a timer on my pump that takes care of when to shut down the mixing.

Once all the methoxide has been introduced into your processor, the color of the solution will fall back to a caramel brown.

Here's the processor hard at work – mixing methoxide and vegetable oil for a minimum of two hours.

Let the solution sit in the processor for a minimum 8 hours

❖ Once the pump has timed out, I'll leave the mixture in my processor for at least 8 hours, and no more than 24 hours. This is important; if you have used sodium hydroxide as your catalyst, and you leave the solution in the processor for too long, the resulting glycerin may get hard and clog up your system. If your system gets clogged, you might consider reengaging the element to get things soft and moving again, but I prefer not to do this given that there's likely unreacted methanol in the water heater.

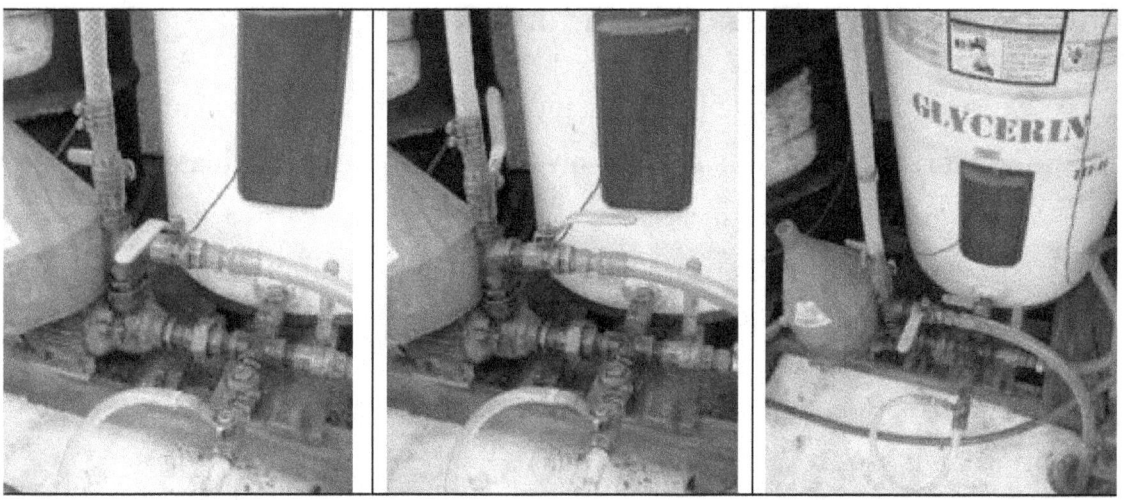

Immediately after turning off the pump, I'll shut off the return flow ball valve at the base of the processor, and then drain the fluid from my sight tube. This prevents biodiesel from degrading the braided tubing, and thus minimizes the risk for a leak.

6.4.8 Drain the glycerin

- ❖ After a minimum of eight hours, I'll return to my processor with containers to capture the glycerin byproduct from the reaction. These containers can be old milk jugs, orange juice containers, or any other disposable container I may have sitting around the house. I've found that the plastic orange juice bottles are the best as far as holding up to chemical attack from residual methanol in the glycerin. Note that plastic milk jugs will leak and/or rupture if not disposed of within a day or two.

Here, I've opened the return valve at the base of my processor as well as the ball valve above my pump. Glycerin will begin to flow immediately and will have the consistency of thin maple syrup.

Here is another view of the glycerin.

- For 35 gallons of oil that requires 1.5 ml of 0.10% sodium hydroxide titrant for a color change, the transesterification of 35 gallons will generally yield about 6 gallons of glycerin. When draining my sixth gallon, I may not immediately see a color change between the glycerin and the biodiesel, but I will see a change in viscosity of flow manifested from slow moving syrupy glycerin to fast moving biodiesel.

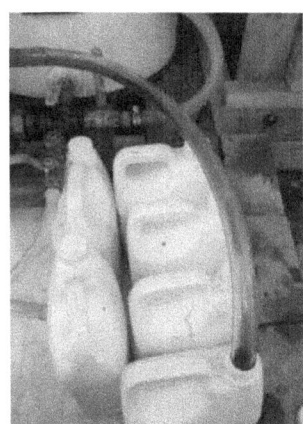

For oil that is mildly acidic, I'll typical generate 6 gallons of glycerin. Upon draining the sixth gallon, I may see a change in viscosity before I see a change in fluid color.

- If the viscosity changes of the drainage drops, but the color of the solution is still relatively dark, I probably have some glycerin not yet separated out of the biodiesel. Rather than introduce this mixture into my washtank, I may collect it in a clear bottle and give that mixture a little extra time to settle and separate into its biodiesel and glycerin phases. I can then pour off the separated biodiesel into my washtank thus helping maintain my yields.

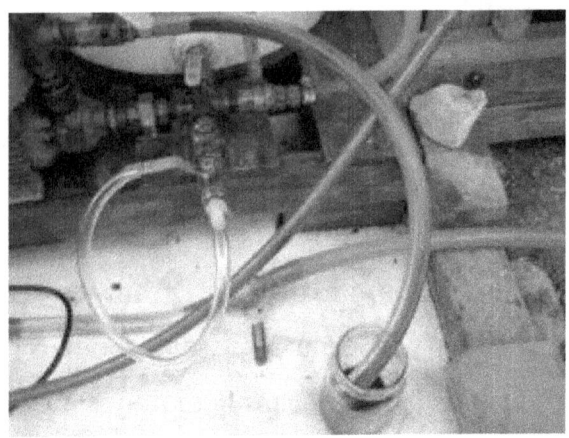

To avoid losing biodiesel with my glycerin, I may capture the low-viscosity dark- colored biodiesel/glycerin mix, and allow it to further separate out. Once it's separated out, I can pour off the biodiesel into my wash tank.

When most of the glycerin from the processor is drained, color alone may not be enough to help determine the breakpoint between glycerin and biodiesel. This is particularly true when using flat-bottomed tanks such as water heaters or barrels for transesterification. As noted in these two pictures, what may appear to be pure glycerin actually contains significant amounts of biodiesel when allowed to settle and separate.

These 7 jars represent the last 7 liters of glycerin collected from one of my biodiesel batches. It's interesting to see how what originally appeared to be glycerin really has significant amounts of biodiesel as evidenced through settling. Using cone-bottom tanks for transesterification helps minimize this issue. Cone bottom tanks provide a smaller surface area to delineate between your biodiesel and glycerin phase during draining.

Use the "Simply Juice" orange juice bottles to drain your last bit of glycerin. These bottles have a nice tapering at the neck makes it simple to drain biodiesel of the top of a glycerin layer.

Over time, suspended biodiesel will start to separate. Note the bottle on the right; it has a different tapering at the neck which makes it more difficult to drain any separated biodiesel.

In about 30 minutes, you should see a clear break between the biodiesel and the glycerin phase.

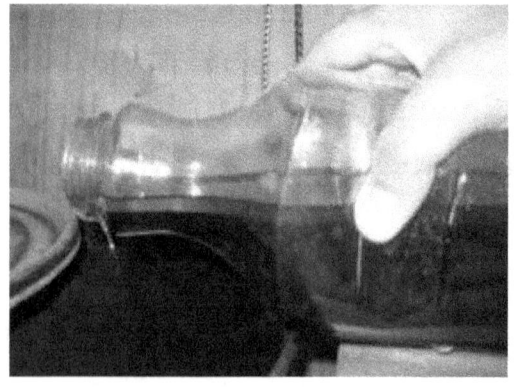

Note how the tapering captures the glycerin and allows you to pour the last bit of separated biodiesel out of the bottle.

Reuse the "Simply Juice" bottle for more separations (or consider buying more juice marketed in these kinds of bottles so that you can do all your separations at once).

6.4.9 Pump the biodiesel into your standpipe wash tank

- Next, I'll connect the delivery port from the processor to my washtank and engage the pump. This will move the biodiesel from the processor into the washtank. I'll get about 33 gallons (+/- two or three gallons) of unwashed biodiesel from the reaction depending on how acidic my oil was. That will fill the 55

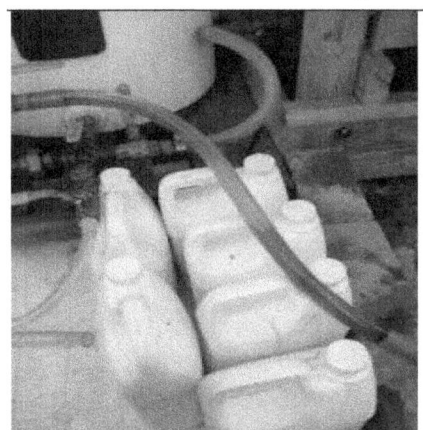 

gallon barrel up to about the second ring on the barrel.

- If you had a good reaction, you will have little residual methanol in your biodiesel. Most of the residual methanol will partition to the glycerin. What little methanol may be found in the biodiesel will either evaporate or readily be washed out with water.

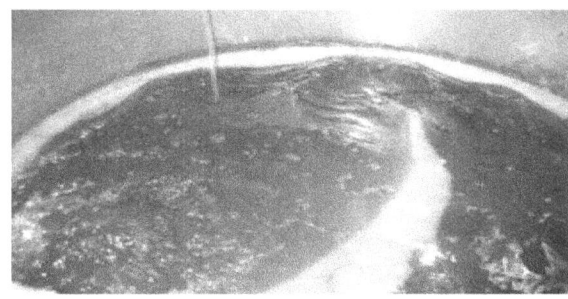

Biodiesel pumped into my wash tank through one of my ball valves. I usually get enough biodiesel to just cover the second ring from the bottom in my 55 gallon drum.

6.4.10 washing biodiesel – Step 1

❖ For my first wash, I take a common garden hose and simply add water to the biodiesel in my wash tank. The water will circulate through the biodiesel and settle to the bottom of the wash tank. This will remove a majority of the soap and/or residual catalyst suspended in my biodiesel.

This represents about 35 gallons of biodiesel in a 55 gallon barrel

For the first wash, I just add water directly to my biodiesel. The resulting mixing will wash out much of the soap. This soap will be entrained in the water phase which will settle out at the base of my barrel.

Biodiesel after adding the water.

❖ I now have a layer of soapy water in bottom 1/3 of my barrel, and wet biodiesel in the upper 2/3 top of my barrel. The biodiesel will take on a cloudy orange appearance given that the soap in the biodiesel is keeping water suspended in the biodiesel. I always notice a big difference between the wash water at the base of my barrel, and what some might claim to be an emulsification at the top of my barrel. On that note, I can say I have never witnessed a serious emulsification from adding water to my wash tank with a hose.

(Note: Some may approach this first wash with much caution; they'll mist their biodiesel with water rather than fill their wash tank with water in order to avoid an emulsification. I used to do this, but found that it was not necessary when dealing with relatively good quality oil (< 3% acidity). Whatever emulsification I may witness from my first wash will be mild and easily "washed out" during subsequent bubble washes.)

- After a minimum two hours, I'll drain the wash water from my barrel. During this first wash, the soap in the fuel will cause the water to adhere to the biodiesel, so it's more difficult to see a break between soapy water and wet biodiesel while draining the tank. As such, I won't rely solely on the appearance of the wash water to determine when to shut off the drain. Instead, I'll gauge how much needs to be drained by watching the level of the mixture in my wash tank, and cutting off the drain as it approaches the second ring on the barrel from the bottom. The second ring represents the proportion of biodiesel originally delivered into my washtank from the processor (about 35 gallons).

Note white shade of wash in tubing from right drain. Wash water shown is captured from a previous washing as demonstrated by closed ball valves. The tubing on the left has biodiesel captured while filling the barrel. There is a ball valve that needs to be opened to release this biodiesel.

I use a pump to keep wash water moving since slow moving soapy water will eventually clog my hose. The green hose transfers soapy water to a drain servicing my home.

This is the appearance of biodiesel after a first wash with a hose. Don't worry if your biodiesel isn't this dark. If there's lots of soap in your biodiesel, chances are good your fuel will be a little lighter in color due to water "sticking" to biodiesel through the action of soap. If you get an emulsification, give your biodiesel a few days to separate from the water. I have never encountered this problem using a hose for the first wash.

I have a bypass valve which I use to occasionally sample wash water for appearance and ph. This is not necessary for a typical operation.

 Jar on right represents wash water collected immediately after adding water to barrel with hose; there is minimal emulsification that separates shortly.

Jar on left represents wash water after two hours of settling and separating in the drum. The pH of this water is about 10.350.

In order to test for the presence of an emulsification, I occasionally collect the last few ounces of drained wash into a jar, and let it sit in the sun for a few days. Although the wash water may appear to represent an emulsification, it rarely yields any significant amounts of biodiesel. The pH of the first wash water is 10.350 for this batch of biodiesel.

6.4.11 washing biodiesel –Step 2

- ❖ For my next wash, I again take a common garden hose and simply add water to the biodiesel in my washtank. The water will circulate through the biodiesel and settle to the bottom of the washtank. As it circulates, it will help remove much of the remaining soap and/or residual catalyst suspended in my biodiesel.
- ❖ Next, I'll bubblewash for two hours. I use a Whisper 400 Aquarium pump with a single wood air-stone. For this round, I don't need to bubblewash any longer than two hours since the soap in my biodiesel will quickly saturate the wash water. After the bubble wash is complete, I'll let my solution rest for at least two hours. The pH of my wash water at this point will be about 10.250. I'll drain this wash water as was done previously.

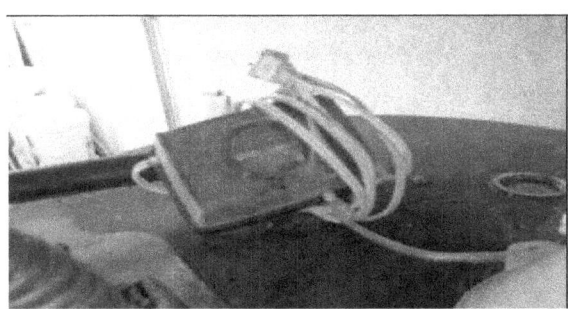
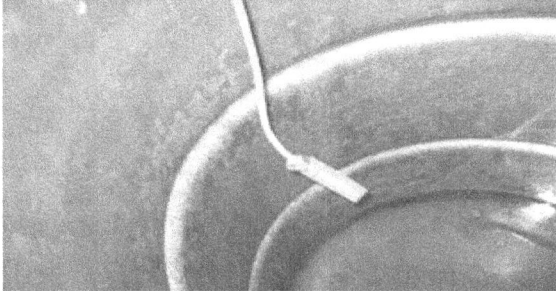

Here are photos of the aquarium pump and wooden air stone used to bubble wash my biodiesel. The air stone normally rests at the base of the barrel under the water layer, and is weighted with washers and or nuts.

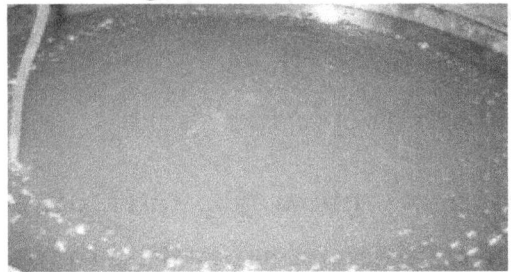

Bubble wash in action

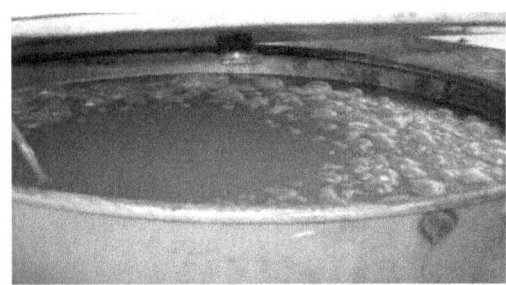

Soap bubbles may precipitate during second and third washes; scoop these out with a fish net to facilitate washing. These will minimize with subsequent washings.

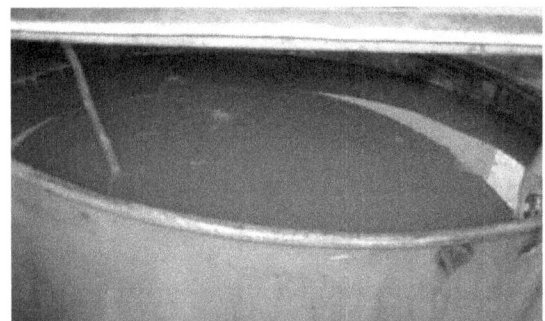

Appearance of biodiesel with most of soap removed after second washing.

Jar on left contains water from first wash and second wash.

PH of second wash is 10.250

6.4.13 washing biodiesel – Step 3

- For my next wash, I again take a common garden hose and simply add water to the biodiesel in my washtank. Again, the water will circulate through the biodiesel and settle to the bottom of the washtank. As it circulates, it will again help remove much of the remaining soap and/or residual catalyst suspended in my biodiesel.

* Next, I'll bubblewash for eight hours. I need to bubblewash a little longer this time to ensure that I saturate my wash water with as much of the soap in my fuel as possible. Once the bubblewash is complete, I'll let the biodiesel sit for at least two hours. The pH of my wash water at this point will be about 8.893. I'll drain as I normally do, although this time, I'll see a much more distinct break between my soapy water and the washed biodiesel.

Initial pH of wash water is 8.6
After 4 hours of bubbling – pH is 8.815
After 6 hours of bubbling – pH is 8.905
After 8 hours of bubbling – pH is 8.983

Appearance of biodiesel after third wash

Jar on left contains water from second wash.

Jar on right contains water from third wash.

PH of third wash is 8.983

6.4.14 washing biodiesel – Step 4

❖ Next, I'll bubblewash for 12 hours. For biodiesel generated from oil with low acidity (i.e. 1.5%), this will be my last wash. The pH of this wash will be slightly higher than tap water (about 7.7) and will not be clear. This is expected since this final wash will have addressed the remaining soap in the biodiesel. The pH of this wash water will be 8.484.

Initial pH of wash water is 8.219
After 10 hours of bubbling – pH is 8.408
After 12 hours of bubbling – pH is 8.484
After 16 hours of bubbling – pH is 8.466

Appearance of biodiesel after fourth wash. It might be a little cloudy if temperatures are cooler outside, and there's no sun. Cloudiness is due to water being in suspension on a cool day.

Appearance of biodiesel after fourth wash.

Jar on left contains water from third wash.

Jar on right contains water from fourthwash.

PH of fourth wash is 8.466

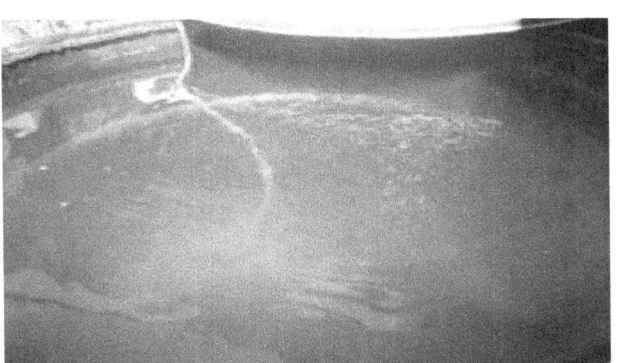 

This is what your final bubblewashed biodiesel might look like outdoors on a warm day sunny. On cooler days, your fuel may not be as transparent as what is shown here since moisture is more likely to be held in the biodiesel phase. Note the misting head in the cover of my barrel (picture on the right). I no longer use this method; bubble washing seems to be adequate enough.

I use a clamp to hold the lid of my barrel open while bubble washing. I've noticed this helps clear up the biodiesel on warm sunny days during washing.

❖ Some guides suggest that you continue washing until the pH of your wash water equals that of tap water, and that the wash water itself be clear. Since I live in a desert, I avoid doing this since 99% of the soap will have been removed by my third bubble wash. If I take a sample of biodiesel from the top of my barrel and mix it with water, it will normally break into a biodiesel/water phase within one minute, and the water phase will be clear. The pH of the wash water in that test will also be much closer to the pH of tap water. By not being married to the condition that my final wash water look like tap water, I can save lots of water. The key is to test the quality of the biodiesel in a separate masonry jar by mixing it with water.

Progression of wash water collectionsduring different periods of the washingprocess for a different batch of biodiesel.

Notice how your wash water will get progressively lighter.

Last jar is tap water.

Here, I've added one part of water to two parts of biodiesel and agitated strongly.

This is the appearance of the fuel water mixture after about 1 minute. The water phase should be very close to that of tap water,

6.4.15 Drain and dry the biodiesel

- ❖ After letting the mixture sit for a minimum 24 hours in the sun, I'll start draining water out of the drain not attached to the standpipe. When I've dropped the level of the biodiesel to the point where I think the standpipe will only be capturing biodiesel, I'll stop draining my water and will take a sample out of the standpipe drain. If that sample is dry (i.e. not significantly cloudy with water), I can begin draining my 55 gallon barrel directly into a separate holding vessel for drying, filtering, and/or fueling my vehicle.

I'm draining water in order to lower the level of my biodiesel to the point where it will intersect the standpipe level of my washtank. The water drain is shown on the left; the biodiesel drain for the standpipe is on the right. There is residual biodiesel from a prior fill in the tubing on the left.

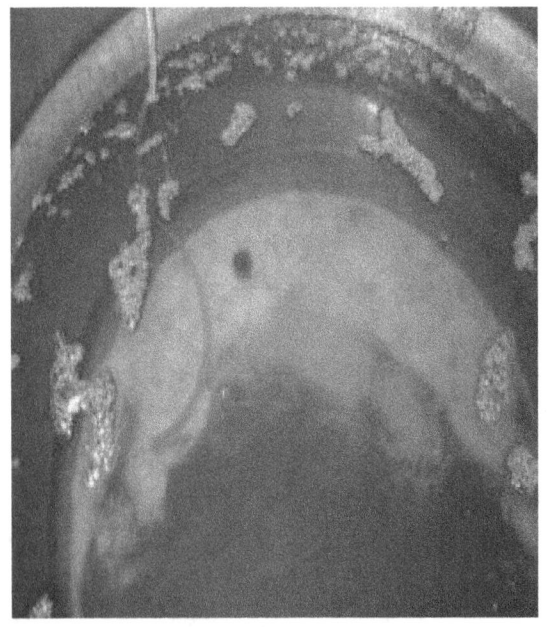

This photo has been taken on a warm sunny day for a batch brewed during the summer. It's easier to see the standpipe through the biodiesel on warm clear days when less moisture is suspended in the biodiesel. If the biodiesel is cloudy, you may have to guess and test as to the level of the water/biodiesel interface as shown in the next two pictures.

Water has been drained to the point where the biodiesel/water interface is now intersecting the level of the standpipe in the washtank. Set the level of the water so that it is slightly below the standpipe so as to avoid introducing water with your drained biodiesel.

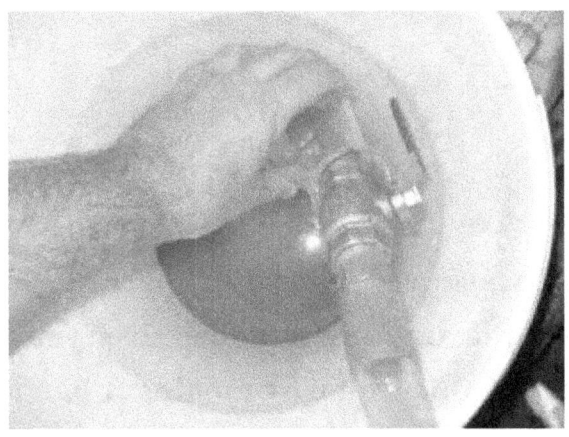 

Here, I am draining biodiesel from the standpipe drain in my washtank into a receptacle. I do this to visually confirm that my standpipe will drain biodiesel rather than a biodiesel / wash water mixture into my storage receptacle. If my biodiesel is clear, the wash water level has been drained sufficiently beneath the standpipe. If my biodiesel is cloudy, either there's water at the level being drained, or my biodiesel has not been given sufficient time to dry.

Here is my biodiesel storage receptacle receiving biodiesel from my standpipe wash tank. This biodiesel was collected on a cool cloudy day and is not as transparent due to water being suspended.

Biodiesel has been drained. Notice how the biodiesel/water interface is slightly below the standpipe. This prevents soaps and/or water to be mixed with my biodiesel during the draining process.

Last of wash water drained via non-standpipe drain. As an alternative, you can recycle this water by making it part of your first wash for your next batch. Also, try not let any biodiesel go down the drain; it can be recycled as part of your next wash.

❖ One way to check for moisture in biodiesel is to insert a 1500 watt heating element into the fuel and check for bubbles. If there is moisture in the fuel, that moisture will bubble off the heating element and rise to the surface. If the fuel is

Photos of my recirculation / dryer tank setup.

crackling and there's a significant amount of bubbles, I'll need to dry my fuel further.

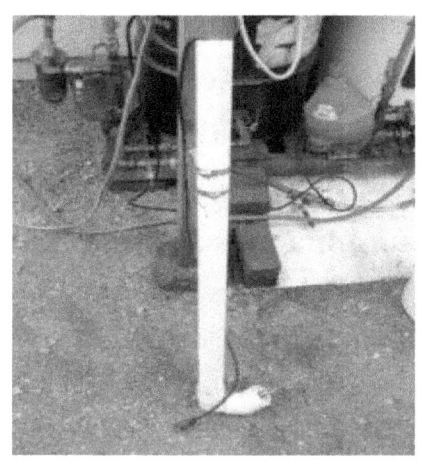
Heating element in PVC tubing for moisture testing. If kept outdoors, this should always be cleaned off of dirt and dust with a fresh disposable rag before inserting in your finished biodiesel. Any remaining dust will be filtered via your fuel filter

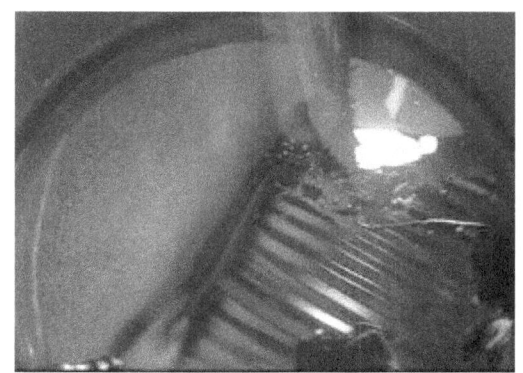
Heating element inserted in biodiesel; note bubbles. If my biodiesel is completely dry, there will only be convection visible above the element.

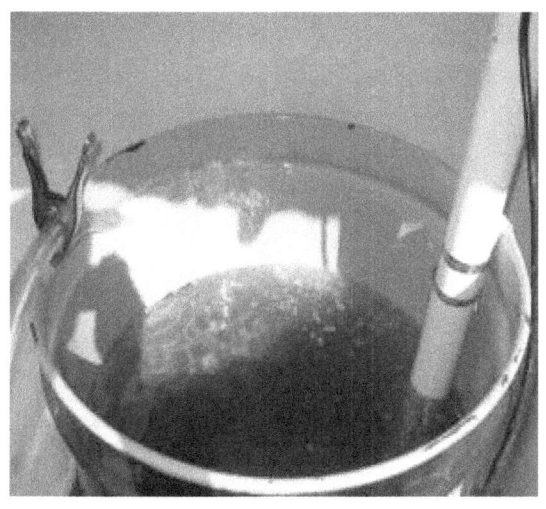
❖ If my biodiesel is still wet, I can speed up the drying process by circulating the biodiesel. The splashing of the biodiesel along the side of my barrel will help evaporate water. Another alternative is to just let it sit and dry in the sun for a few more days. The water will evaporate.

If my biodiesel is still wet, I can speed up the drying process by circulating the biodiesel using an extra 1 inch Clearwater pump. The splashing of the biodiesel along the side of my barrel will help evaporate water. To speed things up even further, I can use my heating element to heat the biodiesel while it is circulating. This requires caution since heating the biodiesel with the element might impact the quality of your fuel, and may also pose a fire risk in the event you have a leak in your system.

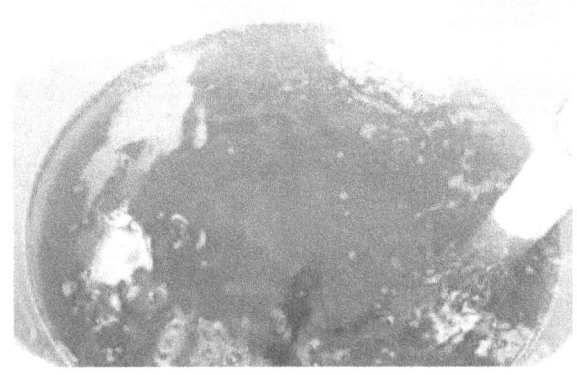

When your biodiesel is dry, you can expect it to be crystal clear and transparent in the sun. On colder days, it may not be as transparent.

6.4.16 Filter the biodiesel

- ❖ I filter my fuel through a standard Goldenrod fuel filter. These will filter down to 10 microns and will also help block any residual water.

This is a standard Goldenrod filter. This will attach to 3/4" pipe. In my setup, the fuel is pumped from my storage tank to a separate holding tank in my garage. These goldenrod filters will filter particles greater than 10 microns in diameter.

- ❖ Finally, I'll polish things off with a five micron whole-house water filter. These are available at Home Depot or Lowes. Note that these do not hold up well to the

sun; if you do this kind of filtering, keep that part of your setup shaded or in the garage

These are photos of fuel storage and delivery system in the garage. My biodiesel is pumped into this tank via the vinyl tubing and elbow attached to the lid. The barrel lid used to be white, but biodiesel is slowly dissolving the paint.

Biodiesel waiting to be consumed; appears dark since photo taken indoors.

This is a picture of a Whirlpool whole house water filter attached to my storage tank. Buy the deluxe model if you want the 5 micron filter.

If you go with this overall setup, make sure the ball valve at the base of your storage barrel is closed to prevent drainage through your delivery hose when not in use.

The ball valve before the filter is redundant if you have one at the base of the barrel. I'm personally okay with redundancy since it minimizes my chances of leaving something open. (It's no fun finding 30 gallons of biodiesel spilled on your concrete garage floor, but it does clean up those old oil drips beautifully.)

When you prepare your filter, make a brace out of 3/4" PVC to help prevent the filter from collapsing under pressure. The packaged filters do not come with this PVC brace shown here inserted into the interior of the filter.

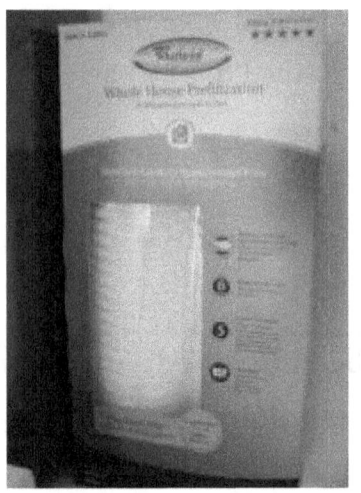

This is the filter to get if you want to filter down to 5 microns. I haven't been able to find a standard fuel filter that will filter down to that particle size. I've filtered over 1200 gallons of biodiesel on just one of these with no issues. They hold up great!

6.4.17 Fill your tank

❖ . ... and here's the best part: fill up your tank and drive off into the sunset knowing that you've helped diminish our reliance on foreign oil, and are truly helping to offset your yearly carbon footprint.

This puts a whole new spin on "fueling at the pump." 20 gallons of biodiesel costs me about $23 to render.

At today's diesel prices, fueling this vehicle with my credit card would require me shell out about $100. Much of this money would probably finance regimes whose policies I don't particularly agree with. It's a lot of work, but it is well worth it!

6.5 Safety issues during processing

Methanol is a poisonous chemical that can cause blindness. The catalyst can cause severe chemical burns if it comes in contact with bare skin. Methanol and lye react to form sodium methoxide (CH3NaO) which is a very caustic chemical. Therefore, always wear proper protective gloves, apron, and eye protection and do not inhale any vapors. Gloves should be chemical-proof with cuffs, it is unsafe to wear shorts or sandals during the handling of any chemicals such as are used for biodiesel production. Always have running water handy when working. The workspace must be thoroughly ventilated. No children or pets should be allowed in the work area.

Avoid exposure to fumes. The greatest danger for fumes is when the methanol is hot. When it's cold or at "room temperature" it fumes very little, if at all, and is easily avoided. Keep methanol at arm's length whenever you open the container. Don't use "open" reactors - biodiesel processors should be closed to the atmosphere, with no fumes escaping. All methanol containers should be kept tightly closed to prevent water absorption from the air.

Transfer methanol by pumping it, with no exposure, which can be achieved using an explosion proof induction pump. Though the mixture gets quite hot at first, no fumes will escape if the container is kept closed. If the methoxide is pumped into a closed biodiesel processor with anexplosion-proof induction pump and is added slowly, which is optimal for the process and also for safety, exposure to fumes will be prevented.

Once again, making biodiesel is safe if you are careful and sensible. "Sensible" also means not overreacting. All chemicals used are common household materials. Lye is sold in supermarkets and hardware stores as a drain-cleaner. Methanol is commonly used in laboratories, and is the main ingredient in the fuel used in racecars. Be careful with these chemicals, but there is no need to be panic regarding the safety.

7.0 References:

- Alovert, Maria. Biodiesel Homebrew Guide: 10th Edition (2005) (unpublished document)
- Blair, Graydon – Utah Biodiesel Supply: Information on rewiring of water heaters:189Hhttp://biodiesel.infopop.cc/eve/forums/a/tpc/f/919605551/m/303 1078611
- Farrel, Alexander et al. Ethanol Can Contribute to Energy and Environmental Goals. Science, Vol 311, January 27, 2006
- Hosein Shapouri et al. The Energy Balance of Corn Ethanol: An Update. U.S. Department of Agriculture, Office of the Chief Economist, Office of Energy Policy and New Uses. Agricultural Economic Report No. 813, 2002
- Pahl, Gregg. Biodiesel: Growing a New Energy Economy. 1st ed. Chelsea Green Publishing Company, White River Junction, Vermont, 2005

- Pimentel, David and Patzek, Tad. Ethanol Production Using Corn, Switchgrass, and Wood;Biodiesel Production Using Soybean and Sunflower, Natural Resources Research, Vol.14, No. 1, March 2005
- Sheehan, John et al. Life Cycle Inventory of Biodiesel and Petroleum Diesel for Use in an UrbanBus. National Renewable Energy Laboratory for the U.S. Department ofEnergy's Office of Fuels Development and U.S. Department of Agriculture's Office ofEnergy, 1998
- Sheehan, John et al. A Look Back at the U.S. Department of Energy's Aquatic Species Program:Biodiesel from Algae, National Renewable Energy Laboratory, 1998
- Tickell, Joshua. From the Fryer to the Fuel Tank: The Complete Guide to Using Vegetable Oilas an Alternative Fuel. 3rd ed. New Orleans: Joshua Tickell Media Productions, 2003
- McCormick, Robert. Effects of Biodiesel on NOx Emissions ARB Biodiesel WorkgroupPresentation; National Renewable Energy Laboratory NREL/PR-540-38296, 2005
- McCormick, Robert et al. Innovation for Our Energy Future: Effects of Biodiesel Blends onVehicle Emissions National Renewable Energy Laboratory Milestone Report NREL/MP540-40554,2006

www.ingramcontent.com/pod-product-compliance
Lightning Source LLC
Chambersburg PA
CBHW082247220526
45469CB00009B/2907